Springer Theses

Recognizing Outstanding Ph.D. Research

Aims and Scope

The series "Springer Theses" brings together a selection of the very best Ph.D. theses from around the world and across the physical sciences. Nominated and endorsed by two recognized specialists, each published volume has been selected for its scientific excellence and the high impact of its contents for the pertinent field of research. For greater accessibility to non-specialists, the published versions include an extended introduction, as well as a foreword by the student's supervisor explaining the special relevance of the work for the field. As a whole, the series will provide a valuable resource both for newcomers to the research fields described, and for other scientists seeking detailed background information on special questions. Finally, it provides an accredited documentation of the valuable contributions made by today's younger generation of scientists.

Theses are accepted into the series by invited nomination only and must fulfill all of the following criteria

- They must be written in good English.
- The topic should fall within the confines of Chemistry, Physics, Earth Sciences, Engineering and related interdisciplinary fields such as Materials, Nanoscience, Chemical Engineering, Complex Systems and Biophysics.
- The work reported in the thesis must represent a significant scientific advance.
- If the thesis includes previously published material, permission to reproduce this must be gained from the respective copyright holder.
- They must have been examined and passed during the 12 months prior to nomination.
- Each thesis should include a foreword by the supervisor outlining the significance of its content.
- The theses should have a clearly defined structure including an introduction accessible to scientists not expert in that particular field.

More information about this series at http://www.springer.com/series/8790

Shiyang Shao

Electrophosphorescent Polymers Based on Polyarylether Hosts

Doctoral Thesis accepted by
University of Chinese Academy of Sciences, China

Springer

Author
Dr. Shiyang Shao
State Key Laboratory of Polymer Physics and Chemistry
Changchun Institute of Applied Chemistry
Chinese Academy of Sciences
Changchun
China

Supervisor
Prof. Lixiang Wang
State Key Laboratory of Polymer Physics and Chemistry
Changchun Institute of Applied Chemistry
Chinese Academy of Sciences
Changchun
China

ISSN 2190-5053 ISSN 2190-5061 (electronic)
ISBN 978-3-662-51513-6 ISBN 978-3-662-44376-7 (eBook)
DOI 10.1007/978-3-662-44376-7

Springer Heidelberg New York Dordrecht London

Softcover reprint of the hardcover 1st edition 2014

Printed on acid-free paper

Springer is part of Springer Science+Business Media (www.springer.com)

Parts of this thesis have been published in the following journals:

1. Shiyang Shao; Junqiao Ding; Lixiang Wang; Xiabin Jing; Fosong Wang. White Electroluminescence from All-Phosphorescent Single Polymers on a Fluorinated Poly(arylene ether phosphine oxide) Backbone Simultaneously Grafted with Blue and Yellow Phosphors. *J. Am. Chem. Soc.* **2012**, *134* (50), 20290.
2. Shiyang Shao; Junqiao Ding; Lixiang Wang; Xiabin Jing; Fosong Wang. Highly Efficient Blue Electrophosphorescent Polymers with Fluorinated Poly(arylene ether phosphine oxide) as Backbone. *J. Am. Chem. Soc.* **2012**, *134* (37), 15189.
3. Shiyang Shao; Zhihua Ma; Junqiao Ding; Lixiang Wang; Xiabin Jing; Fosong Wang. Spiro-Linked Hyperbranched Architecture in Electrophosphorescent Conjugated Polymers for Tailoring Triplet Energy Back Transfer. *Adv. Mater.* **2012**, *24* (15), 2009.
4. Shiyang Shao; Junqiao Ding; Lixiang Wang; Xiabin Jing; Fosong Wang. Synthesis and Characterization of Yellow-Emitting Electrophosphorescent Polymers Based on a Fluorinated Poly(Arylene Ether Phosphine Oxide) Scaffold. *J. Mater. Chem.* **2012**, *22* (47), 24848.
5. Shiyang Shao; Junqiao Ding; Tengling Ye; Zhiyuan Xie; Lixiang Wang; Xiabin Jing; Fosong Wang. A Novel, Bipolar Polymeric Host for Highly Efficient Blue Electrophosphorescence: a Non-Conjugated Poly(arylether) Containing Triphenylphosphine Oxide Units in the Electron-Transporting Main Chain and Carbazole Units in Hole-Transporting Side Chains. *Adv. Mater.* **2011**, *23* (31), 3570

Supervisor's Foreword

It is my pleasure to introduce Dr. Shiyang Shao's original work to be published in the series of *Springer Theses*, which is mainly about how to realize high-performance electrophosphorescent polymers (PhPs) for applications in polymer light-emitting diodes (PLEDs).

Nowadays, PhPs, where the phosphorescent dyes are covalently incorporated into the main chain or side chain of the polymer hosts, have attracted considerable attention since they can not only harvest both singlet and triplet excitons to achieve nearly 100 % internal quantum efficiency, but also be fabricated by low-cost wet processes without the risk of phase segregation observed in the physical blends of polymer hosts and phosphors. In terms of fluorescent system, on the other hand, our group has lots of experience on the design and synthesis of three-primary-color as well as single white polymers based on the conjugated polymeric platform. Therefore, the extension of such conjugated polymers from fluorescence to phosphorescence seems to be a quite normal route of research. With this idea in mind, a series of red and green PhPs that contain polyfluorene and polycarbazole as the main chains, respectively, have been successfully developed. However, we encounter a great challenge in blue and white PhPs because the triplet energy (E_T) of the polymeric backbone is required to be at least 2.75 eV (0.1 eV higher than that of the typical blue phosphor, FIrpic (iridium(III)[bis(4,6-difluorophenyl)-pyridinato-N,C^2]-picolinate) in order to prevent E_T back transfer. In fact, most conjugated polymers have low E_Ts (<2.75 eV), which has also puzzled many research groups worldwide which attempt to design highly efficient blue PhPs.

To solve this problem, the first stage of Dr. Shao's thesis is focused on the design of high E_T polymer hosts. Different from the traditional conjugated counterparts, he presents the invention of a series of nonconjugated polyarylether hosts based on the triphenylphosphine oxide/carbazole hybrid. Owing to the synergistic effect of P=O bond and saturated oxygen atom in the main chain, the E_Ts of these hosts reach up to 2.96 eV, breaking through the limitation of traditional conjugated polymers. The significance of this achievement is that the long-standing problem of E_T back transfer can be effectively overcome in the case of FIrpic-based PLEDs.

As a result, high-performance blue PhPs and all-phosphorescent single white polymers have been achieved with the developed polyarylether as the backbone. It should be noted that, the synthesis of such PhPs is not plain sailing because FIrpic is not compatible with the nucleophilic aromatic substitution polymerization performed at high temperature. To my immense gratification, Dr. Shao has always been a man of persistence when it comes to solving tough problems, and eventually got the desired PhPs. The state-of-the-art efficiencies of these PhPs make me confident that this study will shed light on the further development of high-performance PhPs for displaying and lighting applications.

Additionally, he proposes another strategy to design high-performance PhPs when the E_T of the polymer backbone is close to or even lower than that of the phosphor. A spiro-bridged hyperbranched structure in PhPs is demonstrated to be favorable for the inhibition of both the intramolecular and intermolecular E_T back transfer, thereby leading to the improved device efficiency. This finding provides us a new insight into the relationship between the molecular design and device performance.

Part of this work has been published in high-profile journals such as *J. Am. Chem. Soc* and *Adv. Mater.* I believe that the full publication of this thesis in Springer will further promote the research in the field of material science on PLEDs.

Changchun, May 2014 Prof. Lixiang Wang

Acknowledgments

This thesis study has been an endurance race full of obstacles, but I did not run alone. There are a lot of people to whom I wish to express my gratitude.

First and foremost I would like to thank my supervisor Prof. Lixiang Wang. He has been a fantastic supervisor who contributed his magnificent insight, educational effort, and infinite patience to this thesis, and encouraged me to follow the nature of something unknown, but unusual and groundbreaking. Without his constant support, this thesis would not be possible.

My sincere thanks also go to Dr. Junqiao Ding and Dr. Hui Tong for their generosity in sharing their time and expertise on my thesis. My studies have been carried out much more smoothly, thanks to their support. I am grateful to Prof. Yanxiang Cheng, Prof. Zhiyuan Xie, and Prof. Yanhou Geng, who gave honest criticism and invaluable suggestions to me. I also thank Prof. Yanchun Han, Prof. Dongge Ma, and Prof. Donghang Yan for their priceless collaboration in sharing the experimental instruments and discussing the experimental results. To Miss Guangping Su also goes my heartfelt gratitude for her helpful support.

I am indebted to many group members including Dr. Jun Liu, Dr. Xin Guo, and Dr. Lei Chen who have taught me a lot of experimental skills, and whose achievements have inspired me in these years. I feel fortunate to have been in a research group with so many warm-hearted collaborators with whom I laughed and learned happily, and to all of you I would like to say thank you: Lingcheng Cheng, Ming Wang, Chunlei Bian, Xiaofu Wu, Xiaohui Liu, Bo Chen, Haiyang Song, Hui Li, Bowei Xu, Ying Li, Jing Li, Hongwei Fu, Xuchao Wang, Pengcheng Li, Wei Yuan, Haibo Li, Pan Hu, Yang Wang, Debin Xia… . To all the people whom I worked with but not on the list: thank you for all those good times!

I am also grateful to the 973 Project, the National Natural Science Foundation of China, and the Science Fund for Creative Research Groups for financial support.

Finally, I wish to end this part by heartily acknowledging my sister and my parents for their constant emotional and economic support throughout their lifetimes, without which I could not have reached this stage of my professional life.

Contents

Abbreviations

CIE	Commission Internationale de L'Eclairage
CV	Cyclic voltammetry
DSC	Differential scanning calorimeter
EL	Electroluminescence
E_T	Triplet energy
$(fbi)_2$Iracac	Bis[2-(9,9-diethyl-9*H*-fluoren-2-yl)-1-phenyl-1H-benzimidazolate-C^2,*N*](acetylacetonato)-iridium(III)
FIrpic	Bis[2-(4,6-difluorophenyl)pyridinato-C^2,*N*](picolinato)iridium(III)
HOMO	Highest occupied molecular orbital
IQE	Internal quantum efficiency
ISC	Intersystem crossing
ITO	Indium tin oxide
LUMO	Lowest unoccupied molecular orbital
OLEDs	Organic light emitting diodes
PEDOT: PSS	Poly(3,4-ethylenedioxythiophene): poly(styrenesulfonate)
PhPLEDs	Phosphorescent polymer light emitting diodes
PL	Photoluminescence
PLEDs	Polymer light emitting diodes
PLQY	Photoluminescence quantum yield
PVK	Poly(9-vinylcarbazole)
SWPs	Single white polymers
TEBT	Triplet Energy Back Transfer
TGA	Thermogravimetric analyzer

Chapter 1
General Introduction

Flexible phone screens, luminescent window glass, and large-area lighting roofs, which were incredible about 30 years ago, have now or will in the near future become a part of our lives owing to the emergence of organic light-emitting diodes (OLEDs). Because of their high display quality, low consumption, lightweighting, and flexible features, OLEDs have been recognized as the next-generation display and lighting technology. The research in electroluminescence (EL) from solid organic materials was initiated in 1963, when Pope et al. [1] first observed EL from anthracene crystals. In 1987, Tang et al. at Eastman Kodak Company obtained bright green EL from vacuum-deposited organic films under low driving voltage [2], which had laid the foundation of OLEDs for practical application. The key of their success was that a very thin double-layer film was adopted to confine the excitons in an interface. In 1990, Friend's research group at Cavendish Laboratory of Cambridge University first reported the EL of conjugated polymers (poly(*para*-phenylene vinylene), PPV) [3], which inaugurated the research era of polymer light-emitting diodes (PLEDs). PLEDs are amazing because of their low-cost wet-processing technologies, flexibility, and ease to realize large-area displays and lighting sources.

The commercialization of OLEDs began in 1997, when Pioneer Corporation presented the world's first OLED car audio display. In 1998, Cambridge Display Technology (CDT) Ltd. developed the first monochrome active-matrix PLED display. After 2000, the commercialization of OLEDs accelerated with numerous OLED products developed by many manufacturers, including Pioneer, Sony, Samsung, LG, RITEK (Taiwan), Philips, Dupont, and Dow Chemicals. It is noteworthy that Samsung Mobile Display and LG Display successfully commercialized 55-inch OLED televisions in 2013, despite their relatively high prices.

Although great progress has been achieved for OLEDs, significant hurdles still need to be overcome to realize the mass application of OLEDs. In particular, the development of new materials with high efficiency, high stability, and low cost is of paramount importance, which has long been pursued by both academic and industrial communities.

In this chapter, a brief introduction to the fundamentals of organic electroluminescence is presented first. Then, organic materials used for electroluminescent devices are scanned to give the readers a general impression about material sciences

S. Shao, *Electrophosphorescent Polymers Based on Polyarylether Hosts*,
Springer Theses, DOI 10.1007/978-3-662-44376-7_1

involved in OLEDs. Next, important issues related to this thesis such as triplet energy back transfer, polymer host materials, electrophosphorescent polymer materials are overviewed as the major part. Finally, the design concept of this thesis is outlined.

1.1 Fundamentals of Organic Electroluminescence

Fluorescence and phosphorescence [4]. Fluorescence and phosphorescence are two radiative deactivation pathways of the excited state of organic molecules. The difference between them is that the former involves the singlet excited state, while the latter the triplet excited state. As shown in Jablonski's diagram (Fig. 1.1), a molecule is excited from the ground state (S_0) to a certain singlet excited state (S_n), and then reaches the lowest vibrational state (S_1) via vibrational relaxation or internal conversion. The S_1 state can be dissipated in the form of heat, or radiates a photon (fluorescence) and goes to the S_0 state, or undergoes an intersystem crossing (ISC) process to form a triplet state (T_1) exciton when there is a good orbital coupling between the singlet and triplet states. The radiative transition from the lowest vibrational T_1 state to the S_0 state generates phosphorescence.

As $T_1 \rightarrow S_0$ transition is spin-forbidden, the lifetime of the triplet state ($10^{-6} \sim 10^{-3}$ s) is longer than that of the singlet state ($\sim 10^{-9}$ s). Accordingly, the decay rate of phosphorescence is slower compared with fluorescence. In fact, fluorescence and phosphorescence are competitive processes. Generally, the vibrational relaxation of organic molecules is very quick at room temperature, so most of the excited states reach S_1 state soon after excitation and then goes to the S_0 state. In this case, fluorescence is observed. However, at low temperatures where vibrational relaxation is restrained, intersystem crossing of singlet state to triplet

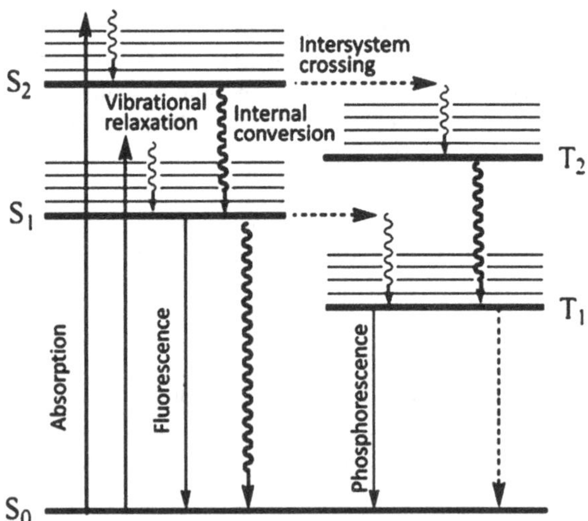

Fig. 1.1 Jablonski diagram

state occurs, leading to much strong phosphorescence from the transition of the T_1 state to S_0 state.

For some transition metal complexes (such as Ir, Pt, Os, and so on), due to the strong spin-orbit coupling induced by heavy metal atoms, the singlet states and triplet states are mixed, making intersystem crossing and the $T_1 \rightarrow S_0$ transition easy to occur. So strong phosphorescence is observed for these complexes even at room temperature [5].

Förster and Dexter energy transfer [6, 7] Förster and Dexter energy transfer are two nonradiative deactivation pathways of the excited state. Förster energy transfer involves the strong dipole–dipole interaction between the energy donor (D*) and acceptor (A). The rate of this process (k_{ET}) is determined by the following equation, where τ_D is the lifetime of the excited state of D*, R is the distance between D and A, and R_0 is the Förster radius:

$$k_{ET}(\text{dipole} - \text{dipole}) = \left(\frac{1}{\tau_D}\right)\left(\frac{R_0}{R}\right)^6$$

Generally, Förster energy transfer is a long-range dipole–dipole interaction (typically in the range of 1–10 nm), and the energy of the singlet/triplet excited state of D can transfer to A via the following processes:

$$^1D^* + {}^1A \xrightarrow{\text{Förster}} {}^1D + {}^1A^*$$

$$^1D^* + {}^3A \xrightarrow{\text{Förster}} {}^1D + {}^3A^*$$

$$^3D^* + {}^1A \xrightarrow{\text{Förster}} {}^1D + {}^1A^*$$

Different from Förster transfer, Dexter energy transfer involves the direct electron exchange between D and A. This process requires a wavefunction overlap between D and A, which means it can only occur at short distance, typically of the order of 15–20 Å. The rate of Dexter energy transfer, k_{ET}(exchange), decays exponentially as the distance between D and A increases, which can be expressed as

$$k_{ET}(\text{exchange}) = KJ\exp\left(-\frac{2R_{DA}}{L}\right)$$

Here K is a constant related to the molecular orbital interactions between D and A. J is a constant related to their spectral overlap. R_{DA} is the distance between them, while L refers to the van der Waals radius. Similar to the Förster energy transfer, the rate of Dexter energy transfer is also proportional to the degree of spectral overlap between the emission spectrum of D and absorption spectrum of A.

Basically, both singlet-singlet and triplet-triplet energy transfer can undergo the Dexter process:

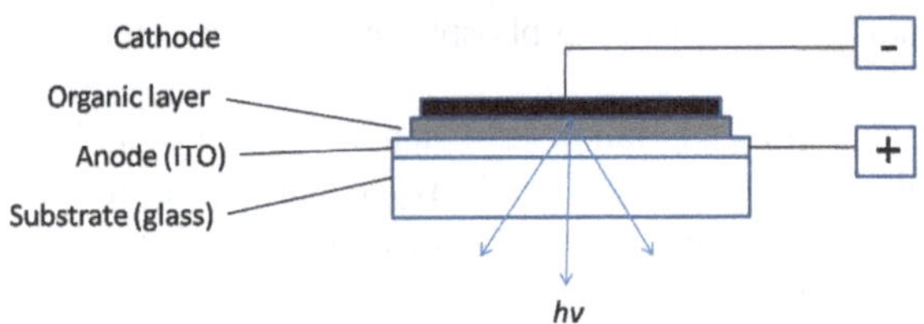

Fig. 1.2 Sandwiched structure of a single-layer OLED

$$^{1}D^{*} + {}^{1}A \xrightarrow{\text{Dexter}} {}^{1}D + {}^{1}A^{*}$$

$$^{3}D^{*} + {}^{1}A \xrightarrow{\text{Dexter}} {}^{1}D + {}^{3}A^{*}$$

Organic light-emitting diodes Organic electroluminescence refers to the phenomena that light emits from organic materials under a certain electric field. Figure 1.2 shows a typical structure of a single-layer OLED, where an organic emissive layer is sandwiched between an anode (e.g., indium tin oxide (ITO)) and a cathode (Al, Mg, Ag, etc.).

OLEDs generally follow the combined theoretical model of inorganic semiconductor devices and molecular orbital theory for organic molecules. As shown in Fig. 1.3, under electric field, electrons and holes are injected into the lowest unoccupied molecular orbital (LUMO) and highest occupied molecular orbital (HOMO) of the organic light-emitting materials from the cathode and the anode, respectively. Then, electrons and holes transport in the organic layer toward the anode and the cathode, respectively. During this process, the recombination of holes and electrons forms excitons. According to the statistical theory, the ratio of singlet and triplet excitons generated during the electroluminescent process is 25 % and 75 %, respectively. For most organic compounds, only the singlet excitons can decay radiatively to generate fluorescence, while all the triplet excitons decay nonradiatively because of the spin-forbidden radiative transition. Therefore, the theoretical internal quantum efficiency (IQE) of the device cannot exceed 25 % for these materials. However, for many transition-metal complexes, due to the spin-orbit coupling effect of the metal atoms, the radiative transition of the triplet exciton becomes allowed.

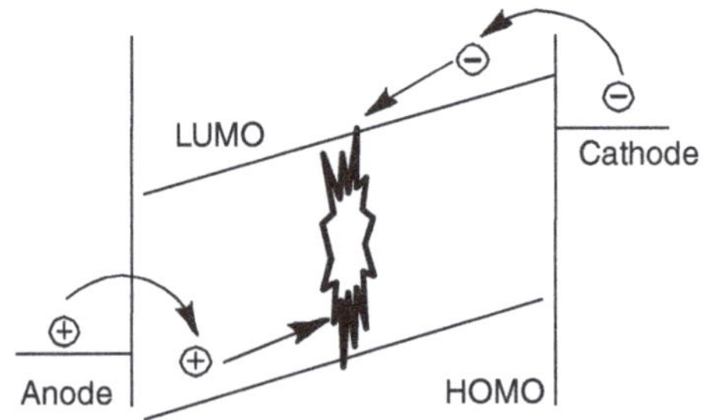

Fig. 1.3 Mechanism of the OLEDs

Consequently, the theoretical IQE of devices based on such materials can reach 100 % [8]. Therefore, development of efficient phosphorescent complexes is an important way to improve the efficiency of electroluminescent devices.

1.2 A Scan of Organic Electroluminescent Materials

Since 1987, great efforts have been paid to organic electroluminescent materials. In order to realize efficient OLEDs, electroluminescent materials should meet the following requirements: (1) balanced electron and hole injection/transporting capability; excess holes or electrons will not only reduce the recombination ratio in the EL process, but also quench the already formed excitons, which leads to low device efficiency; (2) high photoluminescence quantum yield (PLQY) of the emitter which directly affects the radiative efficiency of the excitons; (3) proper energy levels of the materials that match well with the adjacent layers to confine the excitons on the emitters; (4) formation of stable and continuous amorphous film, which is crucial for the fabrication of reliable OLEDs with practical applications. Generally, various materials are needed to simultaneously meet these requirements. According to their functions, they can be divided as luminescent dyes, host materials, carrier transporting materials, interface modification materials, and electrode materials [5, 9–11].

Luminescent dyes Luminescent dyes play the role of emitters in OLEDs. They are excited by means of charge trapping or energy transfer and then produce photons. So their bandgaps directly determine the emission colors of the devices. As the chemical structures of organic molecules are easily tunable, the emission colors of the dyes can be finely adjusted to cover the entire visible range. According to the nature of the excited states, luminescent dyes can be classified as fluorescent and phosphorescent ones. As discussed in Sect. 1.1, for fluorescent dyes, only singlet excitons can be utilized, so the maximum IQE is 25 %, while for phosphorescent ones, both the singlet and triplet excitons can be harvested, so maximum IQE of 100 % can be achieved. Among the various phosphorescent emitters, Ir complexes have been regarded as the most promising candidates because of their stronger phosphorescence at room temperature and shorter emission lifetime relative to most other heavy-metal complexes [5]. Several commonly used fluorescent dyes and phosphorescent ones are listed in Fig. 1.4.

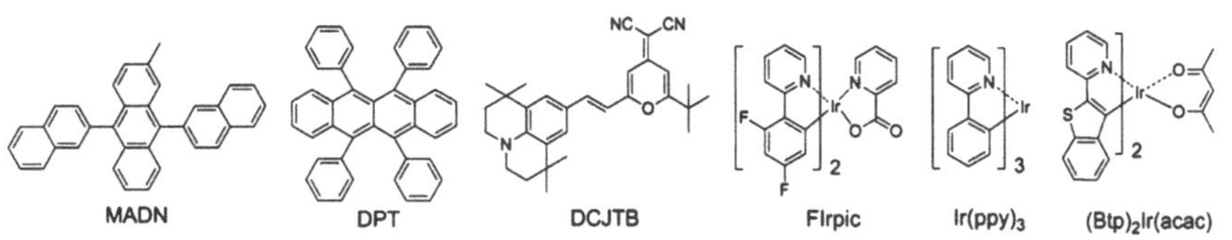

MADN DPT DCJTB FIrpic $Ir(ppy)_3$ $(Btp)_2Ir(acac)$

Fig. 1.4 Chemical structures of several commonly used fluorescent emitters and phosphorescent complexes

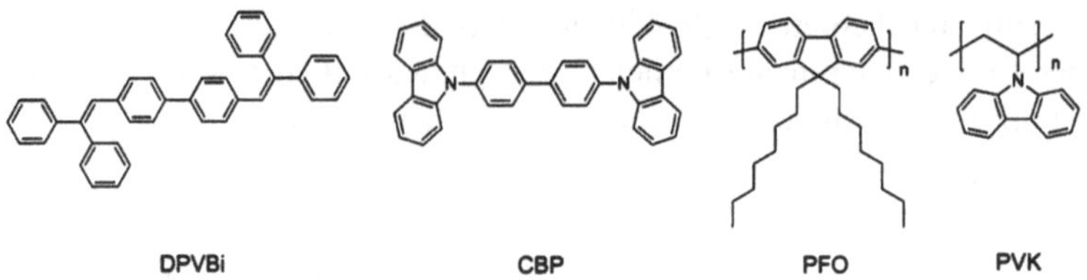

DPVBi CBP PFO PVK

Fig. 1.5 Chemical structures of several hosts

Host materials Since most luminescent dyes show strong concentration quenching effect, they are always dispersed as dopants in suitable host matrix to achieve high device efficiency. Host materials should also be capable for charge injection and transporting in EL process. Moreover, they should also be able to form high-quality amorphous films for the fabrication of reliable OLEDs. So the selection of hosts is critical for OLEDs. Figure 1.5 shows the structures of several host materials.

Carrier transporting materials In order to obtain high device efficiency, the number of electrons and holes in the emissive layer must be balanced. Otherwise, the exciton recombination ratio is low, and more importantly, excess electrons or holes can quench the already formed excitons at high current densities. However, few emissive materials can sufficiently meet this requirement. In order to solve this problem, a number of electron/hole transporting materials have been developed (Fig. 1.6). These materials can be blended with the emissive materials or inserted as an individual layer between the emissive layer and the electrode to achieve carrier balance.

Interface modification materials As OLED devices involve multiple heterogeneous interfaces, the feature of the interface affects the behavior of carriers and excitons and thus considerably determines the device performance. In PLEDs, PEDOT: PSS is mostly used as an anode-modifying layer that can smooth the anode surface and enhance hole injection from anode to the emissive layer. Similarly, PFE [12] and $F(NSO_3)_2$ [13] are examples of cathode-modifying materials, which can enhance electron injection from cathodes to emissive layers (Fig. 1.7).

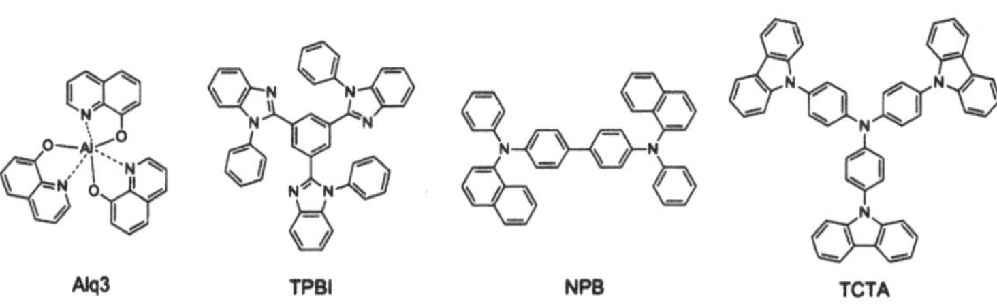

Fig. 1.6 Chemical structures of carrier transporting materials

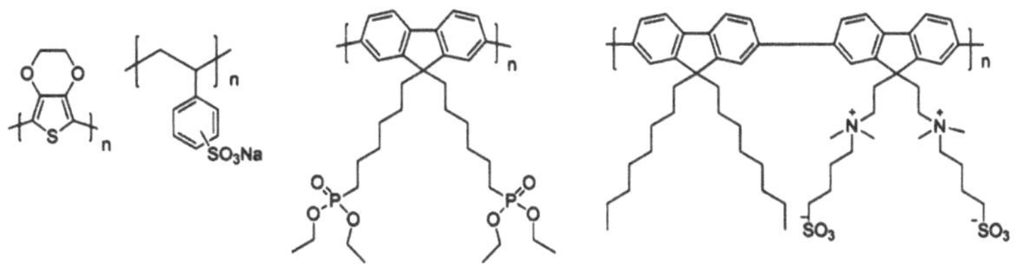

Fig. 1.7 Chemical structures of interface modification materials

Electrodes Generally, the anode materials for OLEDs are metallic oxide with high work function (e.g., indium tin oxide (ITO)), while the cathode materials use low work function metals, such as Ba, Ag, Al, and so on.

According to the processing technology of OLEDs, EL materials can be divided into vacuum deposited ones and wet-processed ones. Generally, low molecular weight molecules are deposited by vacuum deposition method, while polymers with high molecular weights are applied using wet-processed technology, including spin coating, blade coating, and inkjet printing. Although high-quality film could be formed by vacuum deposition, this method suffers from high consumption of energy and materials. In comparison, wet-processed polymers share the advantages of low cost and ease to realize large-area OLEDs.

For emissive polymers, they can be further divided into physical blend systems and single molecule systems according to whether the host materials are chemically bonded with the dopants. Although hosts and dopants can be chosen in a wide range for physical blending systems, these systems suffer from phase separation that affects the efficiency and stability of the devices [14], which is a detrimental problem for the long duration of the devices. Thus, developing single molecule light emitting materials that can effectively inhibit phase separation is amazing.

1.3 Triplet Energy Back Transfer: The Bottleneck in Design of Electrophosphorescent Polymers

Introducing phosphorescent dyes into the emissive layer has been proved to be one of the most important advances in the roadmap of high-efficiency OLEDs. Since phosphorescent emitters generally show relatively long excited-state lifetimes and thus severe concentration quenching effect, they are usually doped in host matrix to achieve high device efficiency. In this case, excitons can be formed on phosphors through three routes: (1) Singlet excitons are formed on the host first and then transferred to the phosphor through Förster/Dexter energy transfer, which is followed by a quick intersystem crossing (ISC) process to form triplet excitons on the phosphor. (2) Triplet excitons are formed on the host through electrical excitation and then undergo forward Dexter transfer to the phosphor. (3) Holes and electrons

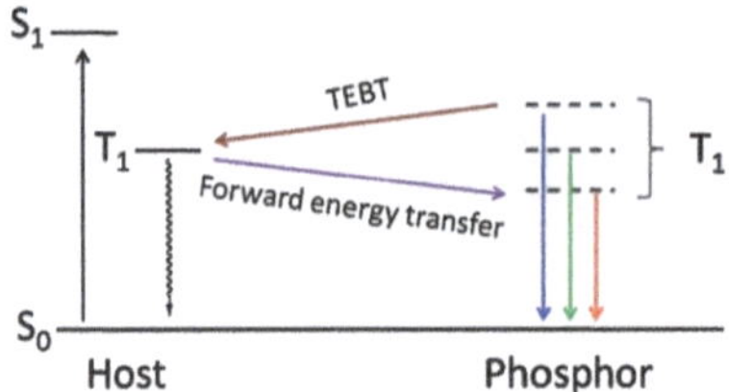

Fig. 1.8 Triplet energy transfer pathways

are trapped by phosphors to directly generate singlet and triplet excitons on phosphors, where the singlets are converted to triplets via ISC process. Generally, forward energy transfer from the host to the emissive phosphor is favorable for EL emission. While the reverse process, i.e., the triplet energy back transfer (TEBT), is unfavorable because the triplet state transition of the host is spin-forbidden and the excitons will be lost in the form of heat (Fig. 1.8). As TEBT of triplet excitons is faster than their radiative decay process, it is easy to occur for phosphors especially when their E_Ts are higher than those of hosts [15], which is rightly the case of most blue phosphorescent PLEDs (PhPLEDs) where the blue phosphors possess high E_Ts (>2.62 eV) but the polymer hosts show low E_Ts (<2.65 eV). Basically, the TEBT problem has become the bottleneck in the development of blue PLEDs as well as the all-phosphorescent white PLEDs.

To inhibit TEBT from phosphors to polymer hosts, there are two main approaches reported so far: (1) elevating E_Ts of polymer hosts, (2) separating phosphors from the hosts.

Elevating E_Ts of polymer hosts. Until now, great efforts have been made by a number of research groups on developing high E_T polymer hosts. These studies include introducing meta-linkage [16–19], or inserting saturated atoms [20, 21] into the polymer mainchain. Unfortunately, for most reported polymer hosts so far, the E_Ts are still lower than 2.65 eV, so they can only host red or green phosphorescent dopants. Developing efficient polymer hosts with E_Ts higher than 2.70 eV has been proved to be a big challenge according to previous reports. Details about these efforts will be discussed in Sect. 1.4.

Separating phosphors from the hosts. In view of the difficulty of developing high E_T polymer hosts, researchers turned to another approach, i.e., separating phosphors from the hosts, to inhibit the TEBT process. As we know, TEBT involves energy transfer from the triplet state of the phosphor to that of the host, which undergoes a Dexter exchange mechanism. Therefore, this process requires a good electron cloud overlap between the pertinent molecular orbitals of the energy donor and the acceptor. As the electron overlap is strongly dependent on the distance between them, TEBT process can be drastically slowed down by separating phosphors from the hosts.

In 2006, Holmes et al. [22] compared two electrophosphorescent polymers with different alkyl spacers ($-C_8H_{16}-$ or none) between the phosphor and the polyfluorene maincain (Fig. 1.9). It was shown that the polymer with a long alkyl spacer

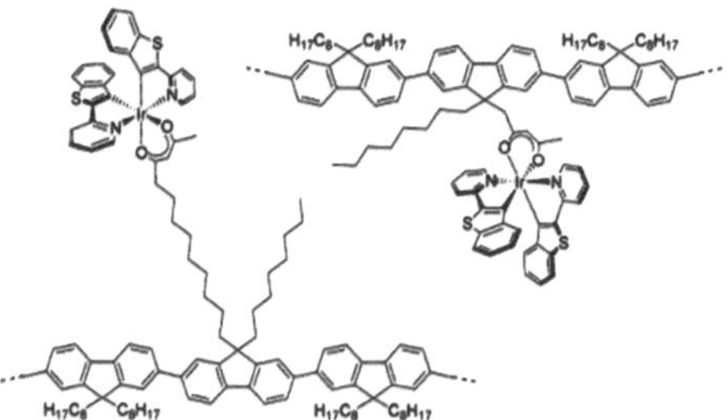

Fig. 1.9 Inhibiting TEBT by introducing a long alkyl spacer between the phosphor and the polyfluorene mainchain reported by Holmes. Reprinted with the permission from Ref. [22]. Copyright 2006 American Chemical Society

shows higher photoluminescence quantum yields (PLQY) and longer phosphorescence lifetimes as well as higher device efficiency compared with that with no spacer. These observations suggested that the TEBT from phosphor to host could be weakened to a certain degree due to the enhancement of the distance between their triplet centers.

Another example was given by Monkman et al. [23] who introduced bulky *tert*-butyl groups on the ligands of a green complex fac-tris(2-phenylpyridine)iridium (III) (1) to obtain **2**, **3**, and **4** (Fig. 1.10). Transient photoinduced absorption measurements of the complexes doped in a low E_T host (poly(9,9'-spirobifluorene), PSBF, $E_T = 2.1$ eV) showed that the build-up rate of triplets on the polymer host followed the sequence of **1** > **4** > **3** > **2**, indicating that the bulky *tert*-butyl groups on the ligands of the Ir complexes inhibited TEBT from the phosphor to the polymer host. In addition, the device using **2** containing two *tert*-butyl groups in each ligand show 3 ~ 6 times improvement of efficiency compared with 3 and **4**

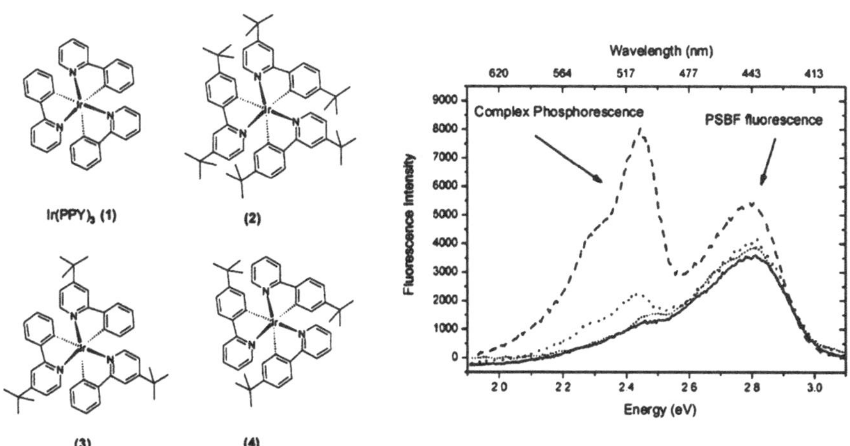

Fig. 1.10 Structures of the Ir complex with different number of *tert*-butyl groups reported by A. P. Monkman. Reproduced from Ref. [23] by permission of John Wiley & Sons Ltd

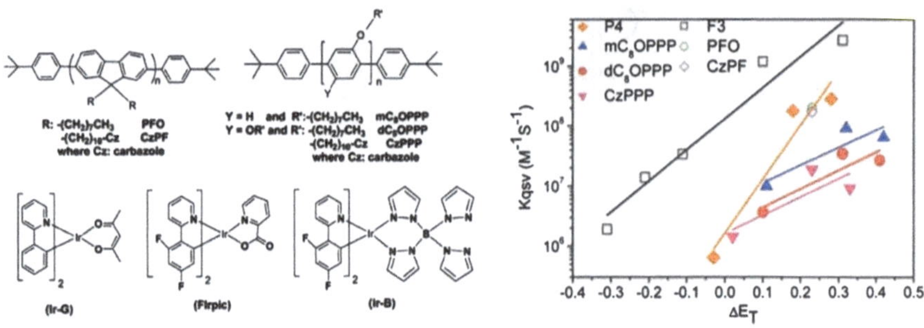

Fig. 1.11 Inhibiting TEBT through the shielding effect. Reprinted with the permission from Ref. [24]. Copyright 2008 American Chemical Society

with only one *tert*-butyl group on the ligand. It is proposed that the bulky groups on the phosphors reduced the electron overlap between the phosphor and the polymer host, and consequently suppressed the TEBT process through Dexter transfer.

Recently, Chen et al. [24] demonstrated that the TEBT process can also be inhibited by the shielding effect of the dense side chain of the polymers. As shown in Fig. 1.11, the Stern–Volmer quenching rate constant (K_{qsv}) is much smaller for the polymers with carbazole side chains than those with alkoxyl or alkyl side chains. Correspondingly, the carbazole-based polymer shows excellent device efficiency of 30 cd A^{-1} for the green electrophosphorescent device, which is about twice that obtained with the alkoxyl based polymer host.

In conclusion, TEBT from phosphors to polymer hosts is a fundamental process that directly affects the EL efficiency of PhPLEDs. Carefully modulating this process is of vital importance for realizing highly device efficiency. Elevating E_Ts of the polymer hosts is the most direct and effective way to inhibit the TEBT process but currently has encountered great challenge, especially for blue PhPLEDs. Great efforts are still needed to search high E_T polymer hosts. Meanwhile, developing ingenious molecular design strategies to inhibit TEBT without using high E_T polymer hosts is also of great significance. However, the reported approaches are basically based on the concept of separating the phosphors from the hosts. Molecular design based on new physical principles to modulate TEBT should be encouraged because it is not only helpful for us to deeply understand the nature of this process, but also will lead to high-efficiency materials with practical application.

1.4 Polymer Host Materials

Suitable polymer host materials are critical for PhPLEDs in terms of efficiency and reliability. A good polymer host should possess two characteristics: suitable E_T and excellent carrier injection/transporting capability.

The prerequisite for a polymer host is that its E_T should be higher than that of the dopant to prevent the triplet exciton loss caused by TEBT (see Sect. 1.3). For red, green, and blue dopants, E_Ts of the hosts should be higher than 2.18 eV, 2.40 eV, and 2.70 eV, respectively. However, for conventional polymer hosts that based on conjugated building blocks, the E_Ts are low because of the large exchange energy, rendering them not suitable for high-energy dopants such as blue ones. So enhancing their E_Ts is necessary. In this section, recent literatures focusing on increasing E_Ts of polymer hosts are reviewed.

In general, decreasing the conjugation length of polymer backbone is the most effective approach to increase E_Ts of conjugated polymers. To reduce conjugation length, *meta*-linkage between the repeating units is a widely used way. In 2004, Addy van Dijken et al. [16] reported a series of carbazole/oxadiazole copolymers (Fig. 1.12). They found that E_T of the copolymer with *meta*-linkage (H1) was higher than that of *para*-connected counterpart (H2). Similarly, Zhang et al. [25] gradually increased the E_Ts of the polycarbazoles from 2.32 eV (H3) to 2.52 eV (H5) by carefully adjusting the connecting position. 4,4′-Bis(*N*-carbazolyl)biphenyl was also polymerized at the 3,6-position of the carbazole units to provide H6 with E_Ts up to 2.53 eV by Chen et al. [26] A luminous efficiency of 23.7 cd A^{-1} is realized for green PhOLEDs based on H6. By polymerizing fluorene units with

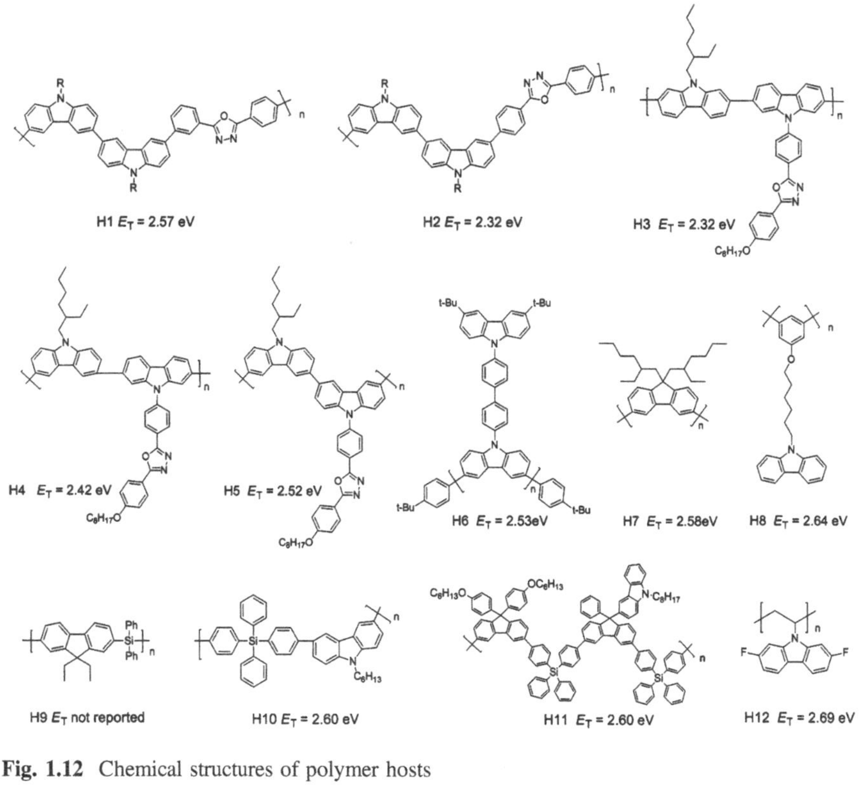

Fig. 1.12 Chemical structures of polymer hosts

meta-linkages, Cao et al. [17] synthesized H7 (E_T = 2.58 eV) whose E_T is 0.4 eV higher than that of poly(2,7-fluorenes). Rencetly, Pei et al. synthesized *meta*-polyphenyls (H8) with E_T of 2.65 eV [19], which had been the highest among conjugated polymer hosts. Blue PhPLEDs based on this host showed a luminous efficiency of 17.9 cd A^{-1} and EQE of 9.3 %.

Another way to reduce conjugation length is to introduce saturated heteroatoms into the polymer backbone. In 2006, Yang et al. [20] first inserted silicon atoms in the main chain of polyfluorene (H9). However, E_T of the host was not reported and the efficiency of the resultant blue device was only 0.6 cd A^{-1}. Subsequently, Ma [21] and Shu et al. [18] also incorporated silicon atoms to the main chain of the conjugated polymers. Combined with the concept of meta-linkage, they obtained H10 and H11 with E_Ts of ~2.60 eV. Blue and green PhPLEDs based on H10 showed device efficiency of 3.4 and 27.6 cd A^{-1}, respectively, while H11 showed a promising luminous efficiency of 32 cd A^{-1} for green PhPLEDs. In fact, the most widely used polymer host PVK and its derivates can be considered as a fully saturated main chain that is expected to possess high E_Ts. Fluorine substituted poly(9-vinylcarbazole) (H12) reported by Takasu et al. showed an E_T of 2.69 eV. Luminous efficiency of 27 cd A^{-1} was obtained with FIrpic as dopant and H12 as the host [27].

Besides high E_T, the polymer host should have excellent carrier injection/transporting capabilities. Currently, the widely used poly(9-vinylcarbazole) (PVK) possesses high E_T and good hole transporting capability. But its HOMO level (−5.9 eV) is too low for hole injection. In addition, its electron transporting ability is significantly lower than the hole transporting ability, which causes the unbalanced electron and hole flow in the emissive layer. Therefore, in most PLEDs, PVK should be blended with a large amount of electron transporting materials, or additional electron transporting layer should be added in the device configuration, which brings about the problems of phase separation or device fabrication complexity [28].

In conclusion, conventional conjugated polymers such as polyfluorene and poly(3,6-carbazole) show low E_{Ts} that can be used only for red and green dopants. On the other hand, nonconjugated hosts such as PVK suffer from charge injection/transporting problems. Therefore, it is challenging but of paramount significance to develop novel polymer host systems with both high E_T and excellent carrier injection/transporting ability.

1.5 Electrophosphorescent Polymers

Electrophosphorescent polymers (PhPs) refer to polymers where the phosphorescent complexes are incorporated into the polymer main chain or side chain via covalent bonds. Because of the strong spin-orbital coupling of the phosphorescent complexes, PhPs can harvest both singlet and triplet excitons to achieve nearly 100 % IQE, which is four times that of fluorescent polymers. In addition, compared with small molecular OLEDs, PhPs-based devices take the advantages of low-cost wet-processing technology and convenience to realize large-area and flexible display devices. Moreover, different from the physically blend systems of phosphorescent dopants and polymer hosts which suffer from potential phase separation [14], PhPs show excellent phase homogeneity which is beneficial for the long duration of the devices. Thus, developing PhPs where phosphorescent dopants are chemically bonded with the polymer hosts to inhibit phase separation is amazing. In this section, recent studies on PhPs are overviewed.

Nonconjugated PhPs Nonconjugated main chain is chosen to construct PhPs because of its relatively high E_T. However, nonconjugated polymers always have poor carrier transporting properties, leading to high driving voltages and low EL efficiencies. The EL efficiencies can be enhanced by introducing suitable charge transporting materials into the polymers by blending or covalently attachment. As early as 2002, Kim et al. first incorporated Ir complex to the side chain of poly (9-vinylcarbazole) using the post-complexation method, and reported the PhPs, poly(Ir(ppy)$_2$(2-(4-vinylphenyl)pyridine))-*co*-vinylcarbazole, NP-1, Fig. 1.13 [29]. EL of these PhPs showed emitting peaks at 512 nm, which are characteristic of the Ir complex. EQE of the PhPs based on a multilayer structure reached 4.4 %. Subsequently, Tokito et al. synthesized red, green, and blue PhPs (NP-2, NP-3, NP-4) by incorporating a small amount (0.1–1.5 mol%) of Ir complex to the PVK side chain through the post-polymerization approach which was more easier to obtain well-defined PhPs. Electron transporting materials were blended with these PhPs to balance the injection/transporting of carriers. PLEDs based on these PhPs and a simple single-layer structure showed EQEs of 5.5 %, 9 %, and 3.5 % for red, green, and blue emissions, respectively [30]. In addition, they found that these values can be further improved to 6.9 %, 11 %, and 6.6 %, respectively, by inserting a hole-blocking layer (aluminiun(III) bis(2-methyl-8-quinolinato) 4-phenylphenolate

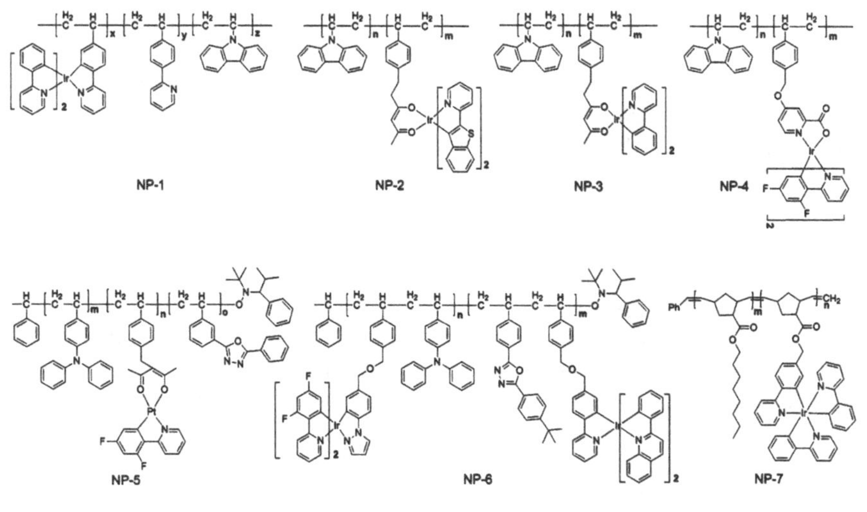

NP-1 NP-2 NP-3 NP-4 NP-5 NP-6 NP-7

Fig. 1.13 Nonconjugated PhPs

(BAlq)) between the emissive layer and the cathode [31]. Fréchet's et al. [32] reported interesting white PhPs by grafting (2-(4',6'-difluorophenyl) pyridinato-*N*, *C*2') (2,4-pentanedionato) Pt(II) unit (FPt), hole-transporting triphenylamine units and electron transporting oxadiazole units to polyethylene chain via the post-complexation method (NP-5). PLEDs based on these PhPs emitted near-white light with CIE coordinates of (0.33, 0.50), which consisted of blue emission from the monomeric FPt moiety and yellow emission from the excimer moiety. Recently, they proposed another approach to achieve white light emission from bichromophoric block copolymers (NP-6) [33]. The micro-phase separation rendered the blue and red dopants separated, thereby inhibiting the energy transfer from blue to red dopants. Finally, by controlling the proportion of blue and red dopants, they realized white electroluminescence from the simultaneous emission of blue and red dopants.

Weck's et al. [34–36] conducted a series of studies on polynorbornene-based PhPs (NP-7). Ir complexes with various emission colors were grafted to the side chain of polynorbornene via the ring-opening metathesis polymerization. All the PhPs exhibited the photophysical properties characteristic of the phosphor analogs, which were found not affected by the inert polymer backbones. Influencing factors of molecular weights of the PhPs, the content of iridium complexes, spacers between the Ir-complex and the mainchain, on the device performance were systematically investigated.

Conjugated PhPs Owing to the good charge transporting ability and proper LUMO/HOMO energy levels for facilitated charge injection, devices based on conjugated PhPs generally show low driving voltages. Nevertheless, E_Ts of most conjugated polymers are rather low, which is insufficient to host high-energy phosphorescent complexes because of the triplet energy back transfer problem. In fact, research on conjugated PhPs has mostly focused on red and white ones until now. There are only few green PhPs and rarely any blue PhP reported to date because of the scarcity of high E_T conjugated polymer hosts.

In 2003, Chen's et al. [37] first reported conjugated PhPs (CP-1) by incorporating $Ir(btp)_2(acac)$ (btp = 2-(2'-benzo[*b*]thienylpyridinato)) and carbazole unit to the side chain of polyfluorenes (Fig. 1.14). The polymers exhibited red electroluminescence from the $Ir(btp)_2(acac)$ units due to the energy transfer from polyfluorene backbone and the carbazole/polyfluorene exciplex to the phosphors. Single-layer device based on these PhPs without injection/transporting layers showed a promising efficiency of 2.8 cd/A.

In 2005, Cao et al. reported a series of red PhPs with poly(carbazole-*alt*-fluorene) as the backbone and red emissive Ir complexes on the side chain (CP2-4) [38]. The poly(carbazole-*alt*-fluorene) mainchain was demonstrated to be a better polymer backbone than polyfluorene because its high HOMO energy level was beneficial for hole injection. Moreover, the *meta*-linkage of the carbazole units could raise the E_T of the polymer backbone and prevent triplet energy back transfer. Single-layer devices of these red PhPs showed high EQE of 4.9 % and luminous efficiency of 4.0 cd/A.

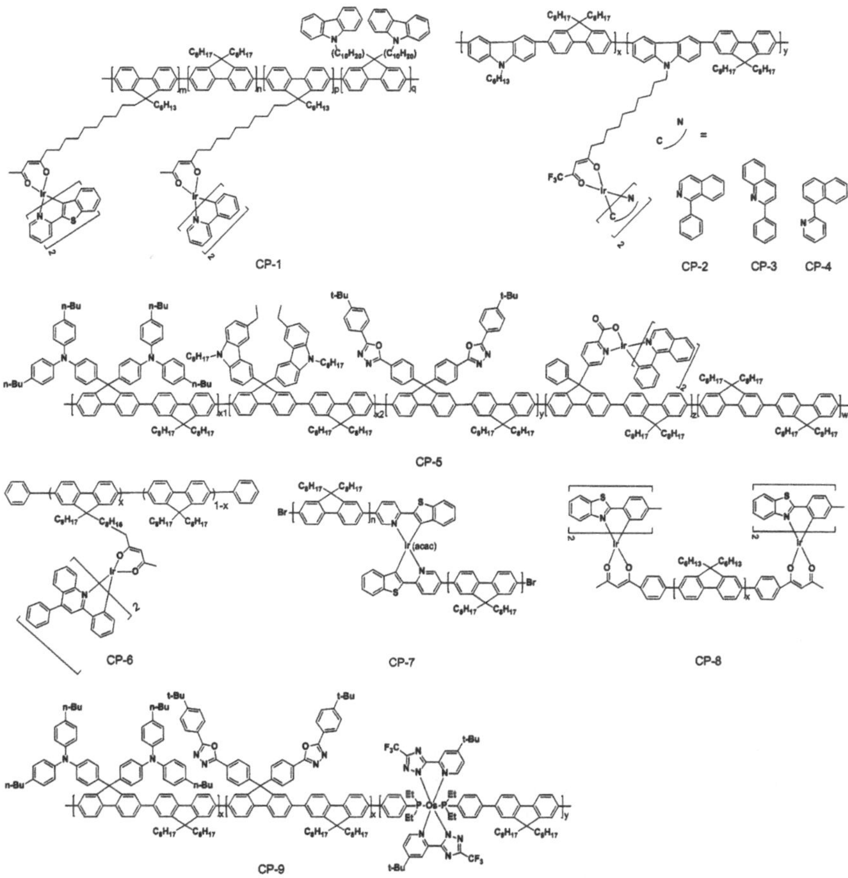

Fig. 1.14 Red light-emitting conjugated PhPs

Shu et al. introduced red-emitting Ir complex to bipolar polyfluorenes with hole- and electron-transporting moieties on the side chain (CP-5) [39]. Excellent charge balance was obtained after optimization of the contents of the hole transporting moieties. Further device optimization led to saturated red PLEDs with luminous efficiency of 9.3 cd/A and power efficiency of 10.5 lm/W.

Recently, Wang et al. reported a red PhP (CP-6) by attaching 1 mol% of Ir $(ppq)_2(acac)$ (ppq = 2,4-diphenylquinolyl-N-$C^{2'}$) to the side chain of polyfluorene [40]. Single-layer device based on this simple polymer exhibited superior EL performance with a luminous efficiency of 5.0 cd/A. Further optimization of the device structure by inserting an alcohol-soluble electron injection layer improved the luminous efficiency to 8.3 cd/A.

Besides red PhPs with phosphors grafted at the side chain, PhPs with phosphors inserted in the mainchain are also developed. Holmes et al. [41] incorporated a series of phosphorescent iridium complexes, including $[Ir(ppy)_2(acac)]$ and

$[Ir(btp)_2(acac)]$ where ppy is 2-phenylpyridinato, btp is 2-(2-benzo[*b*]thienyl)pyridinato and acac is acetylacetonate, into the backbone of oligo(9,9-dioctylfluorenyl-2,7-diyl), where the number of fluorene units reached up to 40 (CP-7). All the PhPs exhibit emission from a mixed triplet state in both PL and EL spectra, with efficient quenching of the oligofluorene singlet emission. Compared with the corresponding complexes, the PhPs showed red-shifted emissions through the oligofluorene attachment owing to the extended conjugation. Yang et al. copolymerized a series of bromimated diketonate anlicary ligand-based Ir complex to the backbone or the terminal of polyfluorenes or poly(fluorene-alt-carbazole)s [42]. It was found that the PhPs with Ir complex on the chain termini (CP-8) showed much higher EL efficiencies than those of the PhPs with Ir complex inserted in the backbone, which was attributed to the suppressed triplet energy back transfer in the former PhPs.

Conjugated PhPs containing complexes of other heavy metals, such as Os, [43] Pt, [44] and Ru [45], have also been developed. Shu et al. reported red PhPs by covalent bonding of a red-emitting osmium complex, into the backbone of a bipolar polyfluorene copolymer (CP-9). PhPs with Os content of 1.5 mol% exhibited red emission at 618 nm with EQE up to 18.0 %, which had been the highest efficiency for red PhPs [43]. The excellent performance was attributed to the balanced charge injections/transporting of the polymer backbone as well as the high phosphorescent efficiency and short triplet lifetime of the Os complex.

Although the polyfluorene skeleton is widely used in red PhPs, it is not suitable for the construction of green PhPs because of its low E_T ($\sim$2.2 eV). Indeed, EL efficiencies of green PhPs are strongly dependent on the E_Ts of polymer hosts. For instance, Wang et al. synthesized a series of green PhPs employing a 1,2-diphenylbenzoimidazole(pbi)-based green Ir complex as the side chain, and poly(3,6-carbazole) (PCz), poly(fluorene-co-3,6-carbazole) (PFCz), and polyfluorene (PF) as the main chain, respectively (CP-10, Fig. 1.15) [46]. Through fine-tuning of the triplet energy of the backbone from 2.2 eV (PF) to 2.6 eV (PCz), the maximum luminous efficiency was improved from 0.3 cd/A to 33.9 cd/A, which was attributed to the efficient triplet exciton confinement on the phosphors.

Compared with red and green PhPs, blue PhPs are rare due to the lack of suitable polymer host with E_T higher than those of the blue complexes (>2.62 eV). Till date, the only known example of blue conjugated PhP is reported by MA et al. who

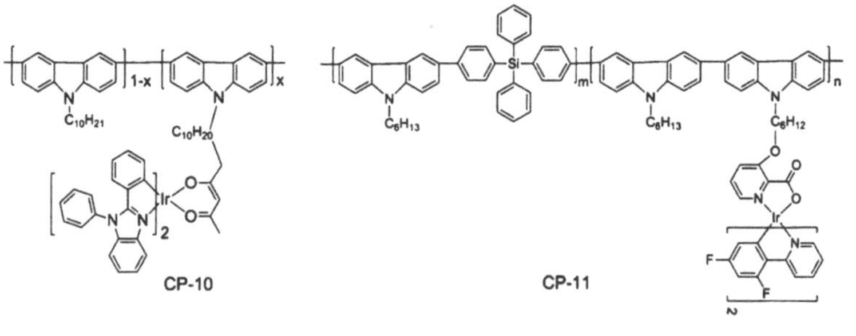

CP-10

CP-11

Fig. 1.15 Green and blue light-emitting conjugated PhPs

grafted 3,6-carbazole-*alt*-tetraphenylsilane copolymers with FIrpic (CP-11) [47]. Although blue emission characteristic of FIrpic was obtained from these polymer films, EL devices based on these PhPs exhibited unsatisfactory luminous efficiency (2.3 cd/A) owing to the low E_T of the polymer backbone ($\sim$2.60 eV).

White light-emitting PhPs and devices have received great attention because of their potential application in lighting sources and backlights. Basically, white PhPs are based on the additive color-mixing theory, where three primary colors (red, green, and blue) or two complementary colors (blue and orange) lead to pure white emission. Owing to the limited suitable polymer host with high E_T for blue and green phosphorescence, only orange and red phosphorescent complexes have been used in white PhPs. Indeed, white emission of these polymers is a mixing of orange/red phosphorescence and blue/green fluorescence. For instance, Cao et al. reported tricolor white PhPs (CP-12, Fig. 1.16) by incorporating a small amount of red emissive Ir complex into the side chain of the fluorene/benzothiadiazole copolymer [48]. By adjusting the contents of the benzothiadiazole unit and the Ir complex to control the energy transfer process, blue and green emission from the mainchain and red emission from the Ir complex could be obtained simultaneously to afford white light. The PhPs showed efficiency as high as 6.1 cd/A, and stable white emission at different applied bias with the CIE coordinates at around (0.32, 0.33). Shu et al. *also* developed white PhPs based on polyfluorenes with benzothiadiazole unit in the main chain and red Ir complex on the side chain [49]. In addition, they incorporated both electron- and hole-transporting units into the polymer to facilitate charge carrier injection and transportation. White devices based on these PhPs exhibited very low turn-on voltage of 2.8 V, and the peak luminous and power efficiencies are 8.2 cd A^{-1} and 7.2 lm W^{-1}, respectively, with the CIE coordinates of (0.33,0.36) at 100 mA cm^{-2}. In the following studies [43], they used an Os complex instead of Ir complex as the red light-emitting species to construct white PhPs (CP-13). Single-layer device based on this polymer showed even higher efficiency of 10.7 cd/A. This improvement was attributed to the higher PLQY and shorter triplet lifetime of the Os complex. Other researchers, including Wang et al. [50], Yang et al. [51], and Shim et al. [52] also reported polyfluorene-based white PhPs. Despite these efforts, EQEs of these white PhPs are still low, which is reasonable considering that the green and blue emitters are still fluorescent ones, thus the excitons are not utilized completely. We believe that designing all-phosphorescent white PhPs where all the emitters are phosphorescent dyes would be an effective way to further increase the device efficiency.

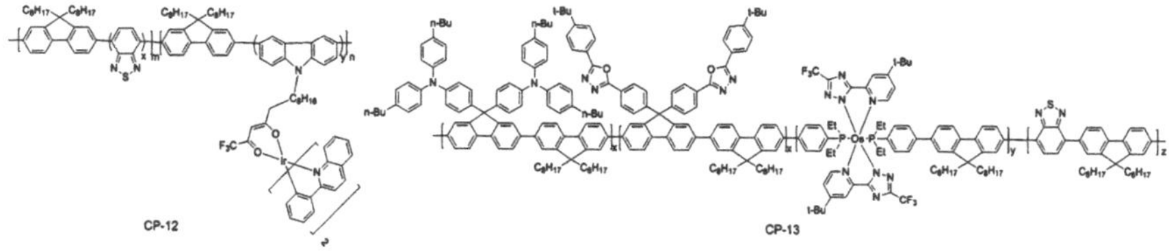

Fig. 1.16 White light-emitting conjugated PhPs

In conclusion, PhPs combine the advantages of high efficiency of phosphorescent complexes and wet processability of polymers. The covalent bond between the phosphors and polymer chains effectively prevents phase aggregation of the phosphors, which not only inhibits the triplet–triplet annihilation between the dopants but also increases the morphological stability of the devices during long-time operation. In the past decade, device performance of PhPs has been considerably improved, indicating that this kind of material is promising in practical application. However, there are still many problems to be overcome. The most urgent one is that blue PhPs are very scarce until now because of the lack of proper polymer host. This problem also brought about the difficulty in designing all-phosphorescent white PhPs. To solve this problem, developing high E_T polymers with excellent carrier injection/transporting capabilities is necessary. Moreover, other issues such as device efficiencies, color stabilities, lifetimes of PhP-based OLEDs also need to be demonstrated.

1.6 Aims of This Thesis

From the above survey, we conclude that the scarcity of high E_T polymer hosts has been a bottleneck in the design of blue PhPs and all-phosphorescent white PhPs. The profound cause of the fact that high E_T polymer hosts should be used lies in the triplet energy back transfer from dopants to hosts.

Therefore, this thesis aims to develop (1) novel polymer host systems that possess both high E_Ts and bipolar injection/transporting capabilities; (2) high-efficiency blue PhPs and all-phosphorescent white PhPs based on the new polymer hosts; (3) alternative approaches to inhibit the undesired TEBT process without using high E_T polymer hosts.

1. Partially conjugated polyarylether hosts are proposed to combine high E_Ts of nonconjugated polymers and excellent charge injection/transport capability of conjugated ones. Their molecular design principles, synthesis, characterization, and device performance are illustrated in detail.

2. Based on the polyarylether hosts, Ir complex FIrpic, is introduced to the polymer side chain to construct blue PhPs. The synthesis and EL performance of this new kind of PhPs is described as the emphasis. In addition, yellow PhPs will also be developed to test the universality of the polyarylether scaffold for other dopants.

3. All-phosphorescent white PhPs will be constructed by simultaneously grafting blue and yellow phosphors to the side chain of the polyether host. The contents of the blue and yellow dopants are adjusted to control the energy transfer process to achieve white emission. Comparison of the all-phosphorescent white PhPs with the traditional white PhPs is also carried out.

4. Inhibiting TEBT based on new physical principles helps us deeply understand the nature of this process. In the last part of this work, a spiro-linked hyperbranched architecture for PhPs is proposed to prevent the TEBT process without high E_T

hosts. The key point of this design is that the vertical conformation between the host and the dopant induced by the spiro-linkage minimized the electron overlap between their pertinent molecular orbitals.

References

1. Pope M, Magnante P, Kallmann HP (1963) Electroluminescence in organic crystals. J Chem Phys 38:2042–2043
2. Tang CW, Vanslyke SA (1987) Organic electroluminescent diodes. Appl Phys Lett 51: 913–915
3. Burroughes JH, Bradley DDC, Brown AR et al (1990) Light-emitting-diodes based on conjugated polymers. Nature 347:539–541
4. Kohler A, Wilson JS, Friend RH (2002) Fluorescence and phosphorescence in organic materials. Adv Eng Mater 4:453–459
5. Evans RC, Douglas P, Winscom CJ (2006) Coordination complexes exhibiting room-temperature phosphorescence: evaluation of their suitability as triplet emitters in organic light emitting diodes. Coord Chem Rev 250:2093–2126
6. Forster T (1959) 10th spiers memorial lecture—transfer mechanisms of electronic excitation. Discuss Faraday Soc 27:7–17
7. Dexter DL (1953) A theory of sensitized luminescence in solids. J Chem Phys 21:836–850
8. Baldo MA, O'Brien DF, You Y et al (1998) Highly efficient phosphorescent emission from organic electroluminescent devices. Nature 395:151–154
9. Xiao LX, Chen ZJ, Qu B et al (2011) Recent progresses on materials for electrophosphorescent organic light-emitting devices. Adv Mater 23:926–952
10. Shirota Y, Kageyama H (2007) Charge carrier transporting molecular materials and their applications in devices. Chem Rev 107:953–1010
11. Tao YT, Yang CL, Qin JG (2011) Organic host materials for phosphorescent organic light-emitting diodes. Chem Soc Rev 40:2943–2970
12. Zhou G, Qian G, Ma L et al (2005) Polyfluorenes with phosphonate groups in the side chains as chemosensors and electroluminescent materials. Macromolecules 38:5416–5424
13. Fang J, Wallikewitz BH, Gao F et al (2011) Conjugated Zwitterionic polyelectrolyte as the charge injection layer for high-performance polymer light-emitting diodes. J Am Chem Soc 133:683–685
14. Noh YY, Lee CL, Kim JJ et al (2003) Energy transfer and device performance in phosphorescent dye doped polymer light emitting diodes. J Chem Phys 118:2853–2864
15. Sudhakar M, Djurovich PI, Hogen-Esch TE et al (2003) Phosphorescence quenching by conjugated polymers. J Am Chem Soc 125:7796–7797
16. van Dijken A, Bastiaansen J, Kiggen NMM et al (2004) Carbazole compounds as host materials for triplet emitters in organic light-emitting diodes: polymer hosts for high-efficiency light-emitting diodes. J Am Chem Soc 126:7718–7727
17. Wu ZL, Xiong Y, Zou JH et al (2008) High-triplet-energy poly 9,9 '-bis(2-ethylihexyl)-3,6-fluorene) as host for blue and green phosphorescent complexes. Adv Mater 20:2359–2364
18. Yeh HC, Chien CH, Shih PI et al (2008) Polymers derived from 3,6-fluorene and tetraphenylsilane derivatives: solution-processable host materials for green phosphorescent oleds. Macromolecules 41:3801–3807
19. Liu J, Pei Q (2010) Poly(M-Phenylene): conjugated polymer host with high triplet energy for efficient blue electrophosphorescence. Macromolecules 43:9608–9612
20. Shao KF, Xu XJ, Yu G et al (2006) Blue electrophosphorescent light-emitting device using a novel nonconjugated polymer as host materials. Chem Lett 35:404–405

21. Fei T, Cheng G, Hu D et al (2009) A wide band gap polymer derived from 3,6-carbazole and tetraphenylsilane as host for green and blue phosphorescent complexes. J Polym Sci, Part A: Polym Chem 47:4784–4792
22. Evans NR, Devi LS, Mak CSK et al (2006) Triplet energy back transfer in conjugated polymers with pendant phosphorescent iridium complexes. J Am Chem Soc 128:6647–6656
23. King SM, Al-Attar HA, Evans RJ et al (2006) The use of substituted Iridium complexes in doped polymer electrophosphorescent devices: the influence of triplet transfer and other factors on enhancing device performance. Adv Funct Mater 16:1043–1050
24. Huang SP, Jen TH, Chen YC et al (2008) Effective shielding of triplet energy transfer to conjugated polymer by its dense side chains from phosphor dopant for highly efficient electrophosphorescence. J Am Chem Soc 130:4699–4707
25. Zhang K, Tao Y, Yang C et al (2008) Synthesis and properties of carbazole main chain copolymers with oxadiazole pendant toward bipolar polymer host: tuning the homo/lumo level and triplet energy. Chem Mater 20:7324–7331
26. Chen YC, Huang GS, Hsiao CC et al (2006) High triplet energy polymer as host for electrophosphorescence with high efficiency. J Am Chem Soc 128:8549–8558
27. Takasu I, Mizuno Y, Uchikoga S et al. (2009) Improvement in triplet exciton confinement of electrophosphorescent device using fluorinated polymer host SPIE:7415, 74150B
28. Mathai MK, Choong V-E, Choulis SA et al (2006) Highly efficient solution processed blue organic electrophosphorescence with 14 Lm/W luminous efficacy. Appl Phys Lett 88:243512
29. Lee CL, Kang NG, Cho YS et al (2003) Polymer electrophosphorescent device: comparison of phosphorescent dye doped and coordinated systems. Opt Mater 21:119–123
30. Tokito S, Suzuki M, Sato F et al (2003) High-efficiency phosphorescent polymer light-emitting devices. Org Electron 4:105–111
31. Tokito S, Suzuki M, Sato F (2003) Improvement of emission efficiency in polymer light-emitting devices based on phosphorescent polymers. Thin Solid Films 445:353–357
32. Furuta PT, Deng L, Garon S et al (2004) Platinum-functionalized random copolymers for use in solution-processible, efficient, near-white organic light-emitting diodes. J Am Chem Soc 126:15388–15389
33. Poulsen DA, Kim BJ, Ma B et al (2010) Site isolation in phosphorescent bichromophoric block copolymers designed for white electroluminescence. Adv Mater 22:77–82
34. Carlise JR, Wang XY, Weck M (2005) Phosphorescent side-chain functionalized poly (norbornene)s containing Iridium complexes. Macromolecules 38:9000–9008
35. Kimyonok A, Domercq B, Haldi A et al (2007) Norbornene-based copolymers with iridium complexes and bis(carbazolyl)fluorene groups in their side-chains and their use in light-emitting diodes. Chem Mater 19:5602–5608
36. Haldi A, Kimyonok A, Domercq B et al (2008) Optimization of orange-emitting electrophosphorescent copolymers for organic light-emitting diodes. Adv Funct Mater 18: 3056–3062
37. Chen XW, Liao JL, Liang YM et al (2003) High-efficiency red-light emission from polyfluorenes grafted with cyclometalated Iridium complexes and charge transport moiety. J Am Chem Soc 125:636–637
38. Jiang JX, Jiang CY, Yang W et al (2005) High-efficiency electrophosphorescent fluorene-alt-carbazole copolymers n-grafted with cyclometalated ir complexes. Macromolecules 38: 4072–4080
39. Yang X-H, Wu F-I, Neher D et al (2008) Efficient red-emitting electrophosphorescent polymers. Chem Mater 20:1629–1635
40. Ma ZH, Ding JQ, Zhang BH et al (2010) Red-emitting polyfluorenes grafted with quinoline-based iridium complex: "simple polymeric chain, unexpected high efficiency". Adv Funct Mater 20:138–146
41. Sandee AJ, Williams CK, Evans NR et al (2004) Solution-processible conjugated electrophosphorescent polymers. J Am Chem Soc 126:7041–7048

42. Zhang K, Chen Z, Yang C et al (2008) Iridium complexes embedded into and end-capped onto phosphorescent polymers: optimizing pled performance and structure-property relationships. J Mater Chem 18:3366–3375
43. Chien CH, Liao SF, Wu CH et al (2008) Electrophosphorescent polyfluorenes containing osmium complexes in the conjugated backbone. Adv Funct Mater 18:1430–1439
44. Zhuang WL, Zhang Y, Hou Q et al (2006) High-efficiency, electrophosphorescent polymers with porphyrin-platinum complexes in the conjugated backbone: synthesis and device performance. J Polym Sci, Part A: Polym Chem 44:4174–4186
45. Wong CT, Chan WK (1999) Yellow light-emitting poly(phenylenevinylene) incorporated with pendant ruthenium bipyridine and terpyridine complexes. Adv Mater 11:455–459
46. Ma Z, Chen L, Ding J et al (2011) Green electrophosphorescent polymers with poly(3,6-carbazole) as the backbone: a linear structure does realize high efficiency. Adv Mater 23:3726–3729
47. Fei T, Cheng G, Hu D et al (2010) Iridium complex grafted to 3,6-carbazole-alt-tetraphenylsilane copolymers for blue electrophosphorescence. J Polym Sci, Part A: Polym Chem 48:1859–1865
48. Jiang JX, Xu YH, Yang W et al (2006) High-efficiency white-light-emitting devices from a single polymer by mixing singlet and triplet emission. Adv Mater 18:1769–1773
49. Wu F-I, Yang X-H, Neher D et al (2007) Efficient white-electrophosphorescent devices based on a single polyfluorene copolymer. Adv Funct Mater 17:1085–1092
50. Mei C, Ding J, Yao B et al (2007) Synthesis and characterization of white-light-emitting polyfluorenes containing orange phosphorescent moieties in the side chain. J Polym Sci, Part A: Polym Chem 45:1746–1757
51. Zhang K, Chen Z, Yang C et al (2008) Stable white electroluminescence from single fluorene-based copolymers: using fluorenone as the green fluorophore and an iridium complex as the red phosphor on the main chain. J Mater Chem 18:291–298
52. Park M-J, Kwak J, Lee J et al (2010) Single chain white-light-emitting polyfluorene copolymers containing iridium complex coordinated on the main chain. Macromolecules 43:1379–1386

Chapter 2
Polyarylether Hosts

2.1 Background

Since phosphorescent emitters generally show relatively long excited-state lifetimes and concentration quenching effect, they are usually doped in suitable hosts to achieve high efficiency [1–3]. So the selection of host materials is critical for fabricating efficient PhOLEDs. Basically, there are several requirements for host materials: (1) E_Ts higher than those of the dopants to prevent TEBT from dopants to hosts (for blue PhOLEDs, $E_T > 2.7$ eV is desirable) [4]; (2) excellent charge injection/transporting capability to achieve considerable charge flow in the emissive layer [5–7]; (3) high thermal and morphological stability to improve operation stability of devices. However, there is generally a dilemma between high E_T and high charge injection/transporting capability for polymer hosts. Polymer with high E_T always possesses high energy bandgap and thus high LUMO level or low HOMO level or both, leading to large charge injection barriers. To solve the problem, great efforts have been paid to developing new polymer host systems, and considerable progress has been achieved for polymer hosts used in green and red PhPLEDs [8–11]. However, polymer hosts suitable for blue PhPLEDs are extremely scarce.

Basically, the ever reported polymer hosts can be classified into two kinds: conjugated ones and nonconjugated ones. Conjugated polymers such as polycarbazoles (PCzs) and polyfluorenes (PFs) usually show low E_Ts (2.4–2.6 eV for PCzs [8] and $\sim$2.2 eV for PFs [9]) because of their large conjugation extent, so they can only host green and red phosphors. To fabricate blue PhPLEDs, the nonconjugated PVK host is most used. However, PVK is a unipolar conductor that transports holes only, so it must be blended with a large amount of electron transporting materials to achieve charge balance [12]. This physical blend system suffers from poor long-term device stability because of phase separation. Therefore, development of efficient novel polymeric hosts for blue PhPLEDs is challenging, but of great significance at present.

S. Shao, *Electrophosphorescent Polymers Based on Polyarylether Hosts*,
Springer Theses, DOI 10.1007/978-3-662-44376-7_2

2.2 Molecular Designs

In this thesis, we report the invention of a series of partially conjugated polyarylether hosts which possess both high E_Ts and bipolarity (Fig. 2.1) [13]. The motivation that we select the polyarylether scaffold is based on the following considerations: (1) the π-conjugation of the polyarylether backbone are interrupted because of the saturated oxygen atom, which will lead to high E_Ts; (2) polyarylethers take advantage of good thermal and morphological stability, which is beneficial for the reliability of the devices; (3) the synthesis of polyarylethers involves a nucleophilic aromatic substitution polycondensation without any residual catalyst contamination which is always a big problem in transition-metal-catalyzed conjugated light-emitting polymers [14]. Based on this polyarylether scaffold, triphenylphosphine oxide and carbazole/triphenylamine units are introduced into the main chain and side chain as electron and hole transport units, respectively, to endow the polymers with bipolar characteristic (PCzPO and PDPAPO, Fig. 2.2). In addition, to investigate the electronic structures of these polyarylethers in detail, their conjugated counterpart (PCzP and PDPAPO) are also designed for comparison.

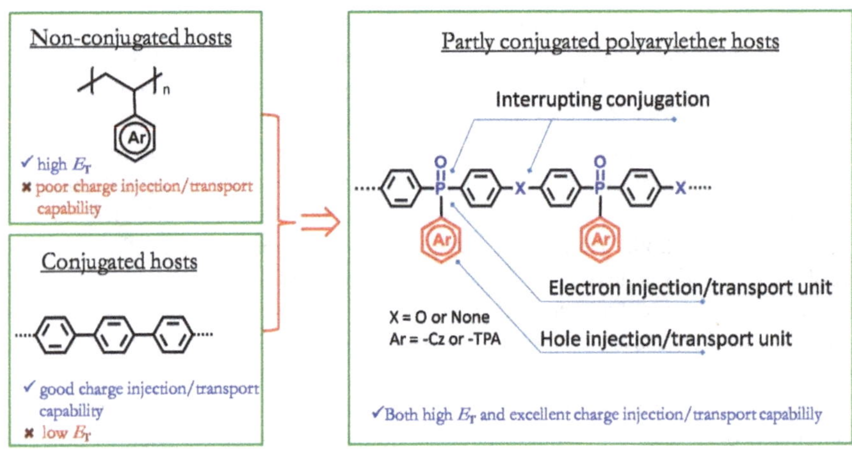

Fig. 2.1 Design strategy of the polyarylether hosts

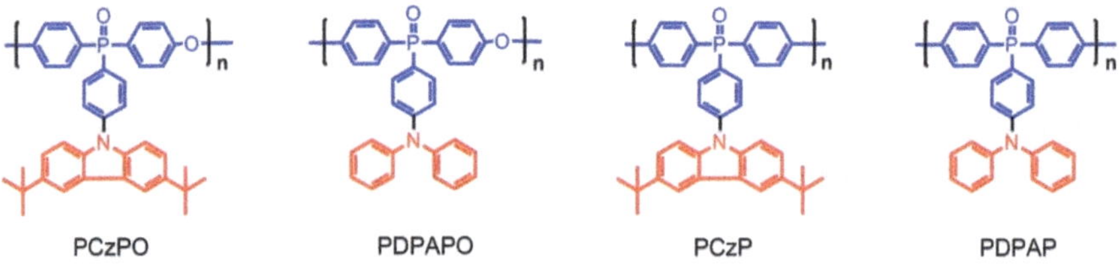

PCzPO PDPAPO PCzP PDPAP

Fig. 2.2 Chemical structures of the polymer hosts

2.3 Results and Discussions

Synthesis and Characterization Scheme 2.1 shows the synthetic route of the polymer hosts. The bisarylphosphine oxide intermediates (**1**, **2**, **3**, **5**) were synthesized from diethylphosphite and corresponding arylmagnesium halides in good yields. The aryl halide intermediates (**6**, **7**, **8**) were produced by Cu-catalyzed *N*-arylization of carbazoles or diphenylamine with *para*-dibromobenzene or *para*-diiodobenzene. The triarylphosphine oxides (F-Cz-MON, F-DPA-MON, Br-Cz-MON, Br-DPA-MON, **9**, **10**) and the model compound (MC) were obtained by Pd-catalyzed coupling reaction between bisarylphosphine oxide and the corresponding aryl halides. In detail, the difluorinated monomers (F-Cz-MON and F-DPA-MON) and alkylsilyloxy precursors (**9**, **10**) can be readily obtained from aryl bromides with corresponding bisarylphosphine oxides. While for the dibrominated monomers (Br-Cz-MON and Br-DPA-MON), aryl iodides were superior in the coupling reaction. Deprotection of the alkylsilyloxy precursors (**9**, **10**) in acid afforded the dihydroxy monomer (HO-Cz-MON and HO-DPA-MON) in nearly quantitative yields. The polyarylether hosts PCzPO and PDPAPO were easily synthesized from the difluorinated monomers and the dihydroxy monomers under classical nucleophilic substitution condition; while the poly(arylene phosphine oxide) type polymer PCzP and PDPAP were synthesized from the dibrominated monomers according to the Yamamoto polymerization with Ni(0) catalyst. All the polymers exhibit good solubility in halogenated solvents such as dichloromethane, chloroform, and chlorobenzene. The number-average molecular weights of the polymers determined by gel permeation chromatography (GPC) with polystyrene as a standard ranges from 9,100 to 20,600 (Table 2.1).

The decomposition temperature (T_d) with 5 % weight-loss was determined by thermogravimetric analysis (TGA). As shown in Fig. 2.3 and Table 2.1, all the polymers show T_ds higher than 440 °C. In addition, the glass transition temperatures (T_g) of the polymers determined by differential scanning calorimetry (DSC) are higher than 200 °C. In special, compared with PCzPO and PDPAPO, PCzP and PDPAP show higher T_gs, indicating that the poly(arylene phosphine oxide) backbone is much more rigid than the oxygen-containing one. Nevertheless, it is noteworthy that T_gs of all the polymers are much higher than the conventional hosts such as polyfluorenes and poly carbazoles. The high T_gs as well as the high T_ds of these polymers endow them with excellent thermal and morphological stability, which is beneficial for the reliability of the light-emitting diodes.

Photophysical Properties The absorption and emission spectra of the polymers are shown in Fig. 2.4. All the polymers show strong absorption in the range of 300–350 nm attributed to the π-π* transition of the repeating units. The PL spectra of the polymers in toluene show similar profiles with peaks in the range of 390–410 nm. While the polymer films showed quite different PL spectra. The emission peaks of the conjugated counterparts (PCzP and PDPAP) were red-shifted by 40–50 nm relative

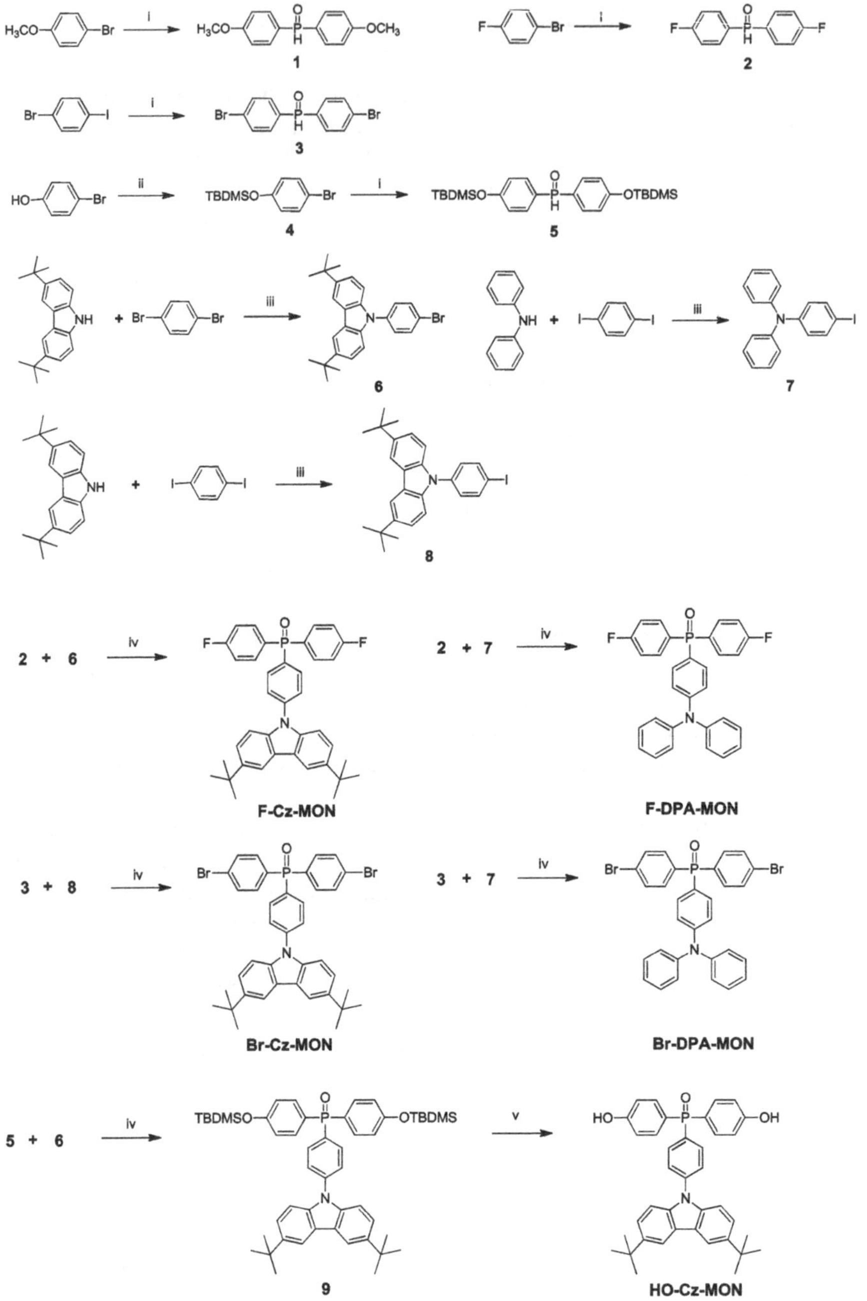

Scheme 2.1 Synthetic route of the polymer hosts. Reagents and conditions: (i) magnesium, THF and diethylphosphite; (ii) *tert*-butyldimethylsilyl chloride, triethylamine and CH_2Cl_2, 0 °C to room temperature; (iii) CuI, K_2CO_3, 1,2-dichlorobenzene and 18-crown-6; 190 °C; (iv) $Pd(PPh_3)_4$, *N*-methylmorpholine and toluene, 100 °C; (v) HCl, room temperature; (vi) BBr_3, CH_2Cl_2, −78 °C to room temperature; (vii) K_2CO_3, *N, N*-dimethylacetamide (DMAc), toluene 170 °C; (viii) Ni $(COD)_2$, COD, 2,2'-bipyridine (BPy), DMF, 80 °C

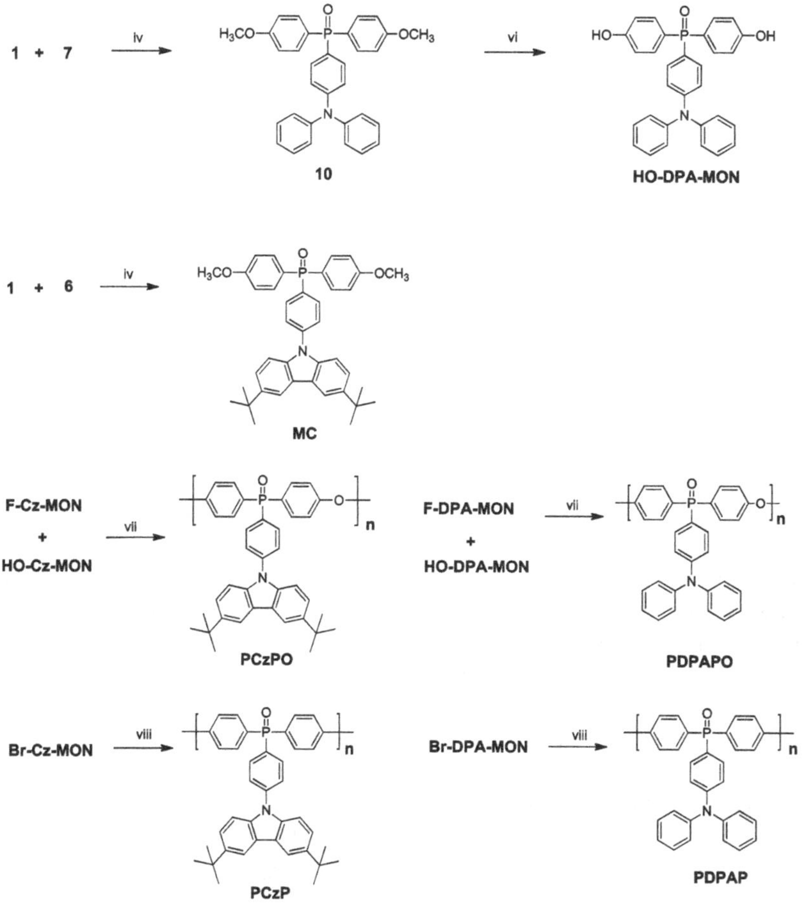

Scheme 2.1 (continued)

Table 2.1 Characterization data of the polymer hosts

Polymer	M_n^a	M_w^b	PDI[c]	T_d^d	T_g^e
PCzPO[f]	20,600	35,600	1.73	513	303
PDPAPO[g]	9,100	20,400	2.24	458	221
PCzP[f]	10,500	17,600	1.68	479	355
PDPAP[g]	10,100	24,100	2.39	448	303

[a] Number-average molecular weight
[b] Weight-average molecular weight
[c] Polydispersity index
[d] Decomposition temperature
[e] Glass transition temperature
[f] GPC performed with THF as the eluent
[g] With $CHCl_3$ as the eluent

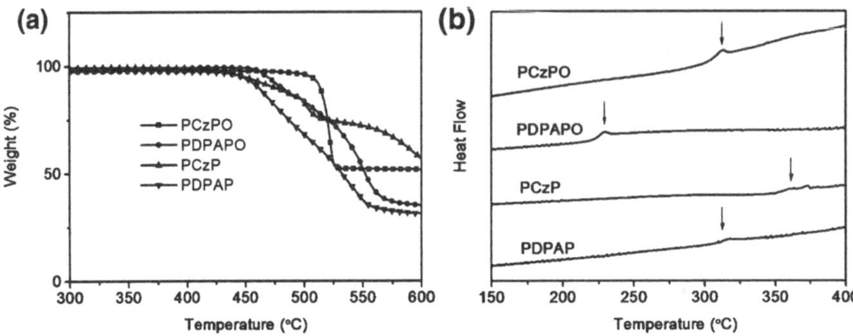

Fig. 2.3 TGA (**a**) and DSC curves (**b**) of the polymer hosts

to those of polyarylethers (PCzPO and PDPAPO). On the other hand, polymers with carbazole side chain (PCzPO and PCzP) showed blue-shifted peaks by 20–35 nm compared with the ones containing diphenylamine side chains (PDPAPO and PDPAP). These observations are consistent with the photo induced charge transfer process from the electron-rich sidechain to the electron-deficient main chain. Polymers with stronger electro-donating units (diphenylamine > carbazole, see below) and/or stronger electro-withdrawing segments (4,4′-bis(diphenylphosphine oxide) biphenyl > triphenylphosphine oxide) show emissions with much longer wavelength.

The phosphorescent spectra of these polymers were measured at 77 K (Fig. 2.5). E_Ts of the polymers were calculated according to the highest-energy triplet vibronic transition. As listed in Table 2.2, all the polymers show E_Ts higher than 2.70 eV, which have broken through the limitation of traditional conjugated polymer hosts. The high E_Ts ensure that these polymers are qualified to host blue phosphorescent emitters such as FIrpic (E_T = 2.62 eV). Considering that the E_Ts of PCzP and PDPAP are much higher than those of polyphenylenes (~2.20 eV), we propose that the introduction of P=O bond into the main chain reduces the conjugation length. As reported in previous studies, P=O bond inserted in aryl units plays the

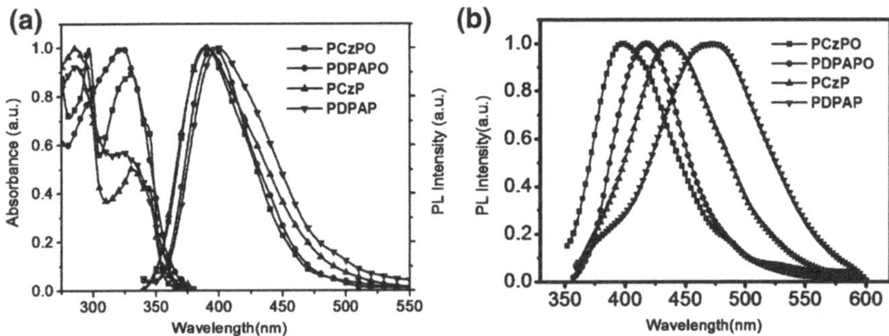

Fig. 2.4 **a** Absorption (in dichloromethane, 10^{-5} mol/L) and PL spectra (in toluene, 10^{-5} mol/L) of the polymer hosts. **b** PL spectra of the polymer films

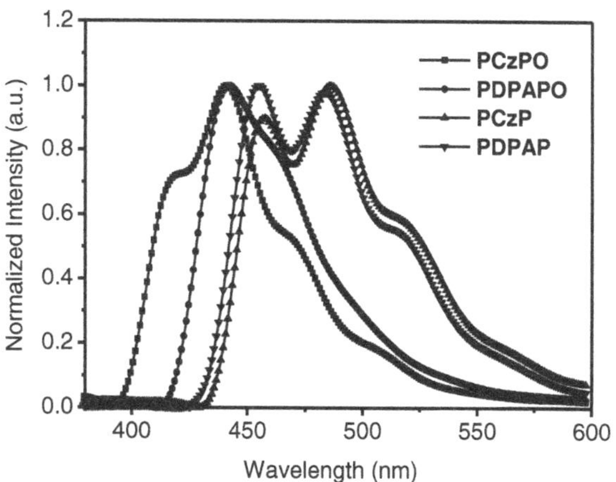

Fig. 2.5 Phosphorescent spectra of the polymer hosts at 77 K (in toluene, 10^{-3} mol/L)

Table 2.2 Photophysical and electrochemical properties of the polymer hosts

	$\lambda^{a}_{em,\ max}$ (nm)	$\lambda^{b}_{em,\ max}$ (nm)	$\lambda^{c}_{abs,\ onset}$ (nm)	E_T (eV)	HOMO (eV)	LUMO (eV)
PCzPO	389	397/417	354	2.96	−5.73	−2.26
PDPAPO	395	418	361	2.81	−5.52	−2.25
PCzP	391	437	354	2.73	−5.71	−2.76
PDPAP	399	466/475	362	2.71	−5.50	−2.78

[a] Emission peaks measured in toluene (10^{-5} mol/L)
[b] Emission peaks measured in film state
[c] Onset value of the absorption spectrum

role of an 'isolated site' so the electrons cannot be delocalized along the adjacent aryl rings [15]. In addition, we note that, compared with PCzP and PDPAP, E_Ts of PCzPO and PDPAPO are elevated by 0.10–0.23 eV. This result indicates that the incorporation of the saturated oxygen atom further interrupts the conjugation of the main chain. Consequently, with the synergistic effect of the oxygen atom and the P=O bond, PCzPO shows the highest E_T of 2.96 eV, which has greatly surpassed most reported polymer hosts in terms of triplet energy (Fig. 2.6).

Electrochemical properties Electrochemical properties of the polymers were investigated by the cyclic voltammetry (CV) (see Fig. 2.7a). The highest occupied molecular orbital (HOMO) and lowest unoccupied molecular orbital (LUMO) energy levels of the polymers are calculated according to their onsets of oxidation and reduction potentials, respectively. As listed in Table 2.2, the HOMO levels of PCzPO and PCzP are −5.73 and −5.71 eV, respectively; while those of PDPAPO and PDPAP were −5.52 and −5.50 eV, respectively. These values are very close to those of their side chain units (carbazole/triphenylamine units). The higher HOMO levels of PDPAPO and PDPAP than those of PCzPO and PCzP are resulted from the stronger electron-donating ability of triphenylamine compared with carbazole

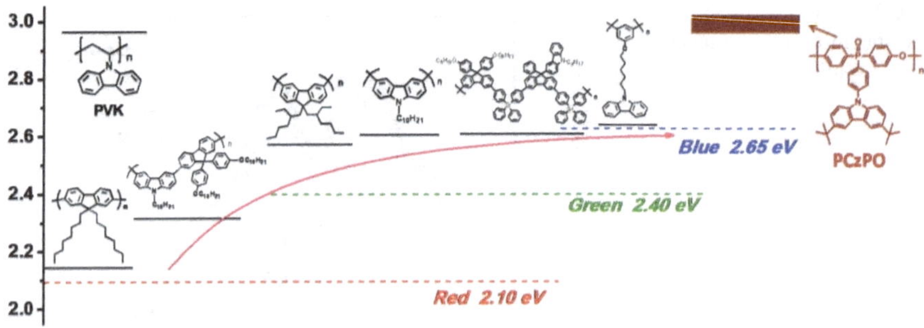

Fig. 2.6 Comparison of E_Ts of PCzPO and traditional polymer hosts

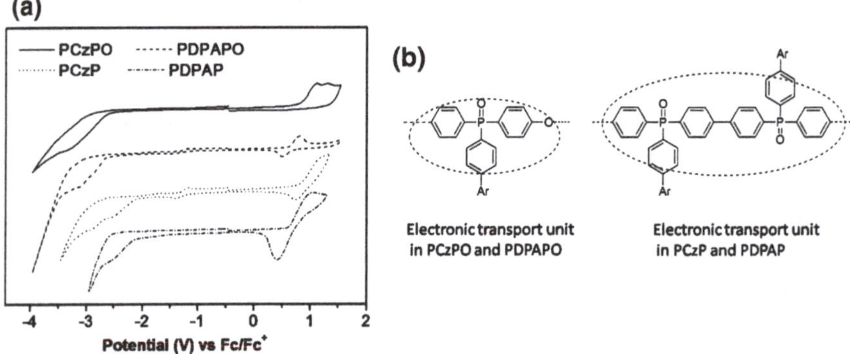

Fig. 2.7 CV curves of the polymer hosts (**a**) and electronic transporting units in the polymer hosts (**b**)

units. The LUMO levels of PCzP and PDPAP calculated from the reduction onsets are at −2.76 and −2.78 eV, respectively, which are about 0.5 eV lower than those of PCzPO and PDPAPO (~−2.25 eV). This observation is reasonable considering that the LUMO energy level is determined by the triarylphosphine oxide segments, therefore the larger conjugation extent caused by the biphenyl units in PCzP and PDPAP mainchian results in lower LUMO level (Fig. 2.7b).

It is worthy to note that, the energy levels of our polymers match better with the common carrier transporting materials and electrodes than those of the widely used host PVK which shows a low HOMO level of −5.90 eV and a high LUMO level of −2.10 eV. The lowered energy barriers between the emissive layer and the adjacent layers are beneficial for the charge injection in the EL process and are expected to enhance the power efficiency of the devices.

Electroluminescent properties In view of the high E_Ts of these polymers, we explore their potential as host materials for blue phosphors. PhPLEDs with typical device architectures of glass/ITO/PEDOT:PSS (40 nm)/polymer hosts: 5 wt% FIrpic (40 nm)/TPCz (50 nm)/LiF (1 nm)/Al (200 nm) were fabricated (Fig. 2.8). Here, TPCz stands for 3,6-bis(diphenylphosphoryl)-9-(4-(diphenylphosphoryl)

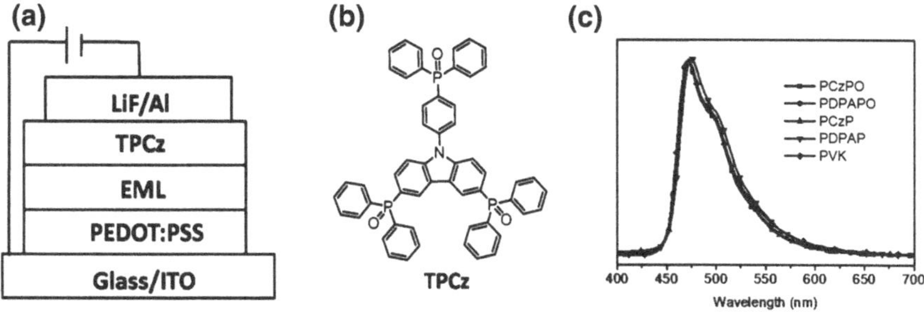

Fig. 2.8 Device configuration (**a**), chemical structure of TPCz (**b**) and EL spectra of the blue devices (**c**)

phenyl)- 9H-carbazole which acts as a hole/exciton blocking material [16]. All the devices show blue EL with a maximum emission peak at 472 nm typical for the emission of FIrpic with no trace of host emission, indicating an efficient energy transfer from the hosts to FIrpic.

The current density (*J*)–voltage (*V*)–brightness (*B*) and luminous efficiency (η_l)–current density (*J*)–power efficiency (η_p) curves are given in Fig. 2.9. The device performance is summarized in Table 2.3. The turn-on voltages of the devices are 5.3 ~ 5.8 V. Devices based on PDPAPO and PCzPO show very promising efficiency with the maximum η_l reaching 23.3 and 20.4 cd A^{-1}, respectively, which correspond to external quantum efficiency (EQE) of 10.8 and 10.0 %, respectively. We believe that both the high E_Ts of PCzPO and PDPAPO and the good carrier injection/transport capability account for the high η_l. In contrast, devices based on PCzP and PDPAP show relatively low η_l of 12.4 and 5.5 cd A^{-1}, respectively. The higher efficiency of PCzPO and PDPAPO than PCzP and PDPAP is reasonable considering that (1) E_Ts of PCzPO and PDPAPO are higher than those of PCzP and PDPAP, which leads to more effective confinement of the triplet excitons on the phosphors; (2) the electron flow in PCzP and PDPAP are comparable with or slightly higher than the hole flow (shown in Fig. 2.10), so considerable electric leakage that leads to the loss of carriers may occur in the PEDOT/(PCzP or PDPAP) interface, which is not prone to occur in PCzPO or PDPAPO-based devices.

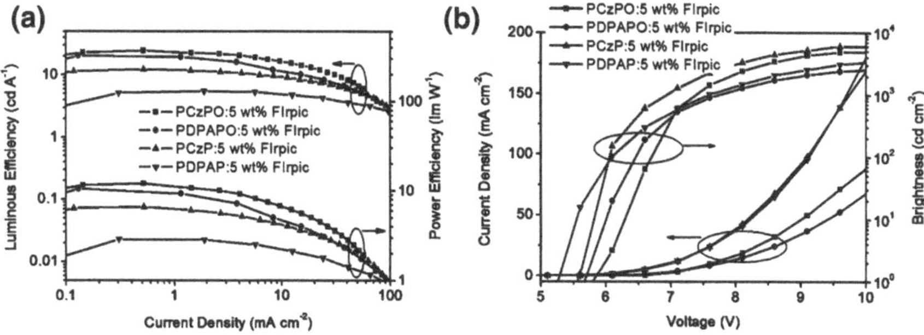

Fig. 2.9 η_l–*J*–η_p (**a**) and *J*–*V*–*B* (**b**) curves of the blue devices

Table 2.3 Device performance of the double-layer blue devices

Polymer	V_{on}^{a} (V)	$\eta_{l,\ max}^{b}$ (cd A^{-1})	$\eta_{p,\ max}^{c}$ (lm W^{-1})	EQE[d] (%)	L_{max}^{e} (cd m^{-2})	CIE[f] (x, y)
PCzPO	5.8	23.3	10.6	10.8	4,860	0.16, 0.30
PDPAPO	5.6	20.4	10.5	10.0	2,603	0.16, 0.31
PCzP	5.5	12.4	6.6	6.1	6,040	0.16, 0.31
PDPAP	5.3	5.5	2.9	2.7	3,484	0.17, 0.34

[a] Turn-on voltage
[b] Maximum luminous efficiency
[c] Maximum power efficiency
[d] Maximum external quantum efficiency
[e] Maximum luminance
[f] CIE coordinates

Since all the polymers contain triarylphosphine oxide units as the electron-transport units and carbazole/triphenylamine units as the hole-transport units, we propose that such polymers should exhibit bipolar characteristics. To confirm this speculation, single-carrier devices were prepared. The hole-only devices contain the following structure: ITO/PEDOT:PSS (50 nm)/Polymer (100 nm)/Au(100 nm), while the electron-only devices contain layers of Al (50 nm)/Polymer (100 nm)/Ca (10 nm)/Al (100 nm). For comparison, control devices using PVK as the active layer were also fabricated.

As Fig. 2.10 shows, PVK shows significantly larger hole flow than electron flow at applied voltage of 7 V ~ 20 V. Whereas in our hosts, the electron flow is significantly enhanced and becomes comparable with the hole flow. Besides, the following points can also be concluded from Fig. 2.10: first, the electron flow of PCzP and PDPAP is larger than that of PCzPO and PDPAPO, which is consistent with the much smaller electron-injection barriers for PCzP and PDPAP induced by their lower LUMO levels; secondly, the electron current of PCzP and PDPAP are in the same order of magnitude as their hole current, which is the foundation for the fabrication of efficient single-layer blue phosphorescent devices as discussed below.

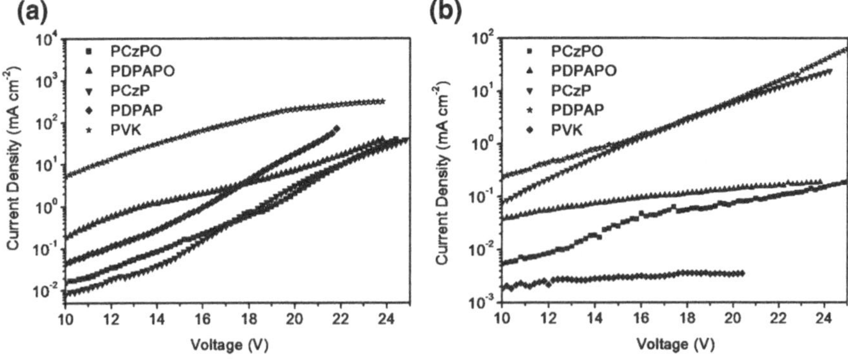

Fig. 2.10 *J–V* curves of the hole- (**a**) and electron-only (**b**) devices

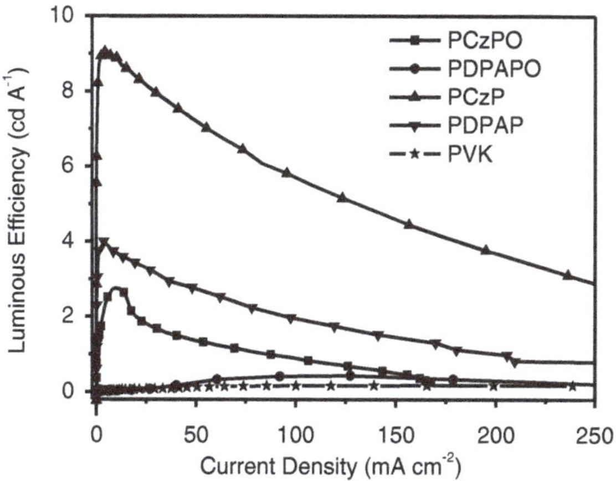

Fig. 2.11 η_l–J curves of the single-layer blue devices

The excellent bipolar characteristic of our polymers is also supported by the performance of the single-layer blue devices with the configuration of ITO/PEDOT: PSS (50 nm)/Polymer: 5 wt% FIrpic/Ca (10 nm)/Al(100 nm).

As shown in Fig. 2.11, all the polymers show higher η_l than PVK under the same current density, implying the much more balanced carriers in the polymers than in PVK. On the other hand, η_l of PCzP and PDPAP is much higher than those of PCzPO and PDPAPO, suggesting that in PCzP and PDPAP, carriers are much more balanced than those in PCzPO and PDPAPO. This observation is in agreement with the single-carrier device results. Among them, devices based on PCzP exhibit the highest luminous efficiency, which reaches 9.0 cd A^{-1}, indicating that PCzP is the most promising host for the fabrication of high-efficiency single-layer devices.

In order to fully exploit the potential of PCzP as a host for single layer devices, we prepared single layer blue, green, and red devices based on PCzP. FIrpic, G1, R1 are selected as the blue, green and red dopant, respectively (see Fig. 2.12). The optimized device structure is glass/ITO/PEDOT: PSS (40 nm)/PCzP: FIrpic or G1 or R1 (100 nm)/CsF (1 nm)/Al (200 nm). Wherein, the optimized doping content of FIrpic, G1 and R1 are 7.5, 30 and 2.5 wt%, respectively. EL spectra of the devices are shown in Fig. 2.13. The emission peaks are located at 476, 540 and 628 nm, corresponding to CIE coordinates of (0.17, 0.35), (0.41, 0.57), and (0.64, 0.35), respectively.

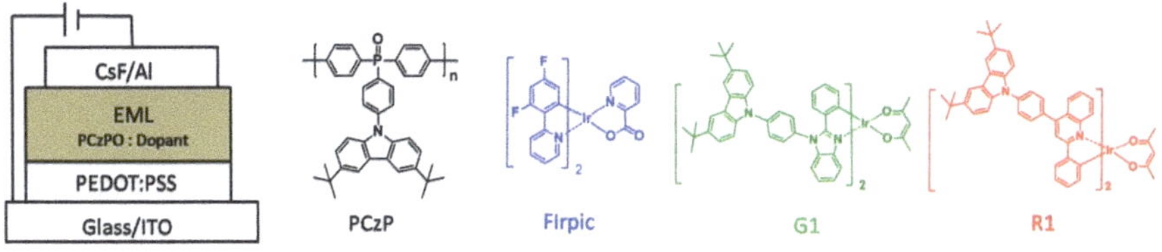

Fig. 2.12 Device structure of the single layer devices as well as the chemical structures of PCzP and the dopants

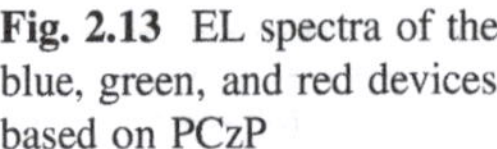

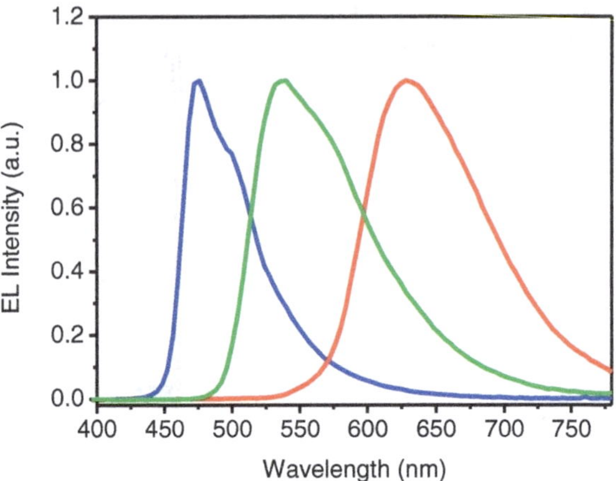

Fig. 2.13 EL spectra of the blue, green, and red devices based on PCzP

As Fig. 2.14 shows, V_{on} of the blue device is 5.7 V, while the maximum η_l, η_p and EQE are 9.8 cd A^{-1}, 4.0 lm W^{-1}, and 4.4 %, respectively. These values, to the best of our knowledge, are the highest ones reported for single-layer blue devices until now. Furthermore, the PCzP-based device shows a very small efficiency roll-off. At a brightness level of 1,000 and 5,000 cd m^{-2}, the luminous efficiency still retain as high as 9.6 and 8.1 cd A^{-1}, respectively. The small efficiency roll-off gives another evidence that the hole and electron flow in the emissive layer is balanced.

The maximum η_l, η_p and EQE of the green device reach 32.3 cd A^{-1}, 21.5 lm W^{-1} and 9.8 %, respectively (Table 2.4). Again the device shows a very small efficiency roll-off with the luminous efficiency retain as high as 31.9 and 27.5 cd A^{-1} at a brightness level of 1,000 and 5,000 cd m^{-2}, respectively. We note that these efficiencies are even comparable with those obtained with many double-layer devices [8–10, 17], indicating that the bipolar PCzP is competent for both the hole and electron injection/transport. Red device with 2.5 wt% R1 as the dopant shows maximum η_l, η_p and EQE of 4.9 cd A^{-1}, 1.4 lm W^{-1}, and 4.9 %, respectively. At a brightness level of 1,000 and 5,000 cd m^{-2}, the luminous efficiency

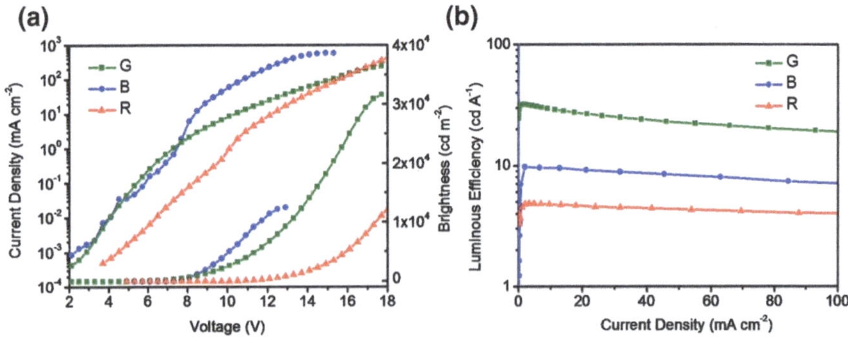

Fig. 2.14 *J–V–B* (**a**) and η_l–*J* curves (**b**) of the blue, green and red single layer devices

Table 2.4 Device performance of the blue, green, and red single-layer devices

Device	V_{on} (V)	$\eta_{l,\ max}$ (cd A^{-1})	$\eta_{p,\ max}$ (lm W^{-1})	EQE (%)	L_{max} (cd m^{-2})	CIE (x, y)
7.5 wt% Firpic: PCzPO	5.7	9.8	4.0	4.4	12,500	0.17, 0.35
30 wt% G1: PCzPO	3.7	32.3	21.5	9.8	31,675	0.41, 0.57
2.5 wt% R1: PCzPO	6.5	4.9	1.4	4.9	13,686	0.64, 0.35

retains as high as 4.6 and 3.9 cd A^{-1}, respectively. These values are also among the highest ones ever reported for red PhPLEDs. These results are encouraging considering that PCzP is so far the only one which can be used as the host material for all the blue, green, and red single layer devices. We believe that excellent performance based on PCzP comes from its high triplet energy levels and its good bipolar characteristic.

In conclusion, we have designed and synthesized a series of bipolar, high triplet energy polyarylether hosts containing triphenylphosphine oxide units as electron-transport main chain and carbazole/triphenylamine units as hole-transport side chain. The polymers show E_Ts as high as 2.96 eV because of the interrupted conjugation induced by the synergistic effect of P = O bond and saturated oxygen atom in the main chain. This achievement has broken through the limitation of traditional conjugated polymer hosts in terms of triplet energy. Blue PhPLEDs based on PCzPO with a double layer configuration exhibits a very promising luminous efficiency of 23.3 cd A^{-1}, which is among the highest ones ever reported for blue PhPLEDs.

Besides high E_Ts, the polymers show excellent bipolar characteristic compared with the traditional PVK host. Because of the simultaneous incorporation of hole and electron injection/transport units, the polymers show much lower energy barriers for the carrier injection, and much more balanced carrier flow compared with PVK. As a result, all the polymers show much lower driving voltage and higher efficiency for the simple single-layer blue device. Especially, PCzP with the most balanced carrier flow shows maximum luminous efficiency of 9.8, 32.3, and 4.8 cd A^{-1} for blue, green, and red single-layer devices, respectively. This result is encouraging considering that PCzP is so far the only one which can be used as the host material for all blue, green, and red single-layer devices.

The significance of this work is that it provides the only known bipolar, high triplet energy polymer host system for blue PhPLEDs. We believe that our study will push forward the development of the PhPLEDs in such a way that the molecular design strategy in this work enlightens researchers to develop rich and varied polymer hosts for PhPLEDs, especially for blue ones. In addition, the importance of these polymers is further emphasized by the fact that they afford a unique platform for the construction of electrophosphorescent polymers which have been recognized as a very promising kind of materials for fabricating efficient, stable, and low-cost PhPLEDs.

2.4 Experimental Section

General Information All chemicals and reagents were used as received from commercial sources without further purification. Solvents for chemical synthesis were purified according to the standard procedures. ^{1}H NMR spectra were recorded with a Bruker Avance 300 NMR spectrometer. Elemental analysis was performed using a Bio-Rad elemental analysis system. MALDI-TOF mass spectra were performed on an AXIMA CFR MS apparatus (COMPACT). TPCz and FIrpic were prepared in our lab.

Measurements and Characterization Molecular weight of the polymers were determined by gel permeation chromatography (GPC) on a Waters 410 instrument with polystyrene as the standard and THF/$CHCl_3$ as the eluent. Thermal properties of the polymers were analyzed with a Perkin-Elmer-TGA 7 instrument under nitrogen at a heating rate of 10 °C/min. UV–visible (UV–vis) absorption and photoluminescent spectra were measured with a Perkin-Elmer Lambda 35 UV–vis spectrometer and a Perkin-Elmer LS 50B spectrofluorometer, respectively. Phosphorescent spectra at 77 K were measured in toluene. Cyclic Voltammetry experiments were performed on an EG&G 283 (Princeton Applied Research) potentiostat/galvanostat system. The samples were tested in acetonitrile using ferrocene as an internal reference and n-Bu_4NClO_4 as the supporting electrolyte. The HOMO energy levels were calculated according to the equation $E_{HOMO} = -e[E_{onset,\ ox} + 4.8\ V]$, in which $E_{onset,\ ox}$ was the onset of the oxidation potential. The LUMO energy levels were calculated according to the equation $E_{LUMO} = -e[E_{onset,\ red} + 4.8\ V]$, where $E_{onset,\ red}$ was the onset of the reduction potential.

PLED Fabrication and Measurements To fabricate PLEDs, a 40-nm-thick PEDOT:PSS film was first deposited on the pre-cleaned ITO-glass substrates (20Ω per square) and subsequently baked at 120 °C for 40 min. For single-layer blue PhPLEDs, FIrpic was doped into the hosts (PCzPO or PVK) and spin-coated onto PEDOT:PSS as the emissive layer (EML). The thickness of the EML was about 100 nm. Successively, a 10-nm-thick film of Ca (or 1 nm CsF), and a 200-nm-thick film of Al were thermally evaporated on top of the EML at a base pressure less than 10^{-6} Torr (1 Torr = 133.32 Pa) through a shadow mask with an array of 10 mm^2 openings. The double-layer device was fabricated by the same method except that a 50-nm-thick film of TPCz was evaporated before deposition of 1 nm LiF and 200 nm Al. Hole-only and electron-only devices were fabricated according to the similar procedure with the structure of glass/ITO/PEDOT:PSS (40 nm)/PCzPO (130 nm)/Au (60 nm) and glass/PEDOT:PSS (40 nm)/Al (100 nm)/PCzPO (130 nm)/Ca (10 nm)/Al (200 nm), respectively. The EL spectra and CIE coordinates were measured using a PR650 spectra colorimeter. The current–voltage and brightness–voltage curves of devices were measured using a Keithley 2400/2000 source meter and a calibrated silicon photodiode. All the measurements were carried out at room temperature under ambient conditions.

Synthesis **1**. Diethylphosphite (0.91 mL, 7.08 mmol) was added dropwise at 0 °C to a solution of 4-methoxyphenylmagnesium bromide in tetrahydrofuran which was prepared from 4-bromoanisole (3 mL, 23.4 mmol) and magnesium (0.674 g, 28.1 mmol). The mixture was aged for 30 min at 0 °C, then stirred at ambient temperature for 16 h. After that it was cooled again to 0 °C, and 75 mL NH_4Cl aqueous was then added slowly. The mixture was extracted with diethyl ether and the organic phase was washed with brine, then it was dried over Na_2SO_4. After the solvent had been completely removed, the residue was purified by column chromatography on silica gel using petroleum ether/ethyl acetate = 1/1 as eluent to give the product in 75 % yield. ^{1}H NMR (300 MHz, $CDCl_3$, δ): 8.03 (d, J = 477.6 Hz, 1H), 7.61 (dd, J = 13.1, 8.5 Hz, 4H), 6.99 (dd, J = 8.6, 1.8 Hz, 4H), 3.85 (s, 6H).

2. This compound was prepared from 4-fluorophenylmagnesium bromide and diethylphosphite according to the procedure for the synthesis of **1**. Column chromatography on silica gel using petroleum ether/ethyl acetate = 1/1 as eluent gave the product as a white solid in 72 % yield. ^{1}HNMR (300 MHz, $CDCl_3$, δ): 8.09 (d, J = 467.7 Hz, 1H), 7.75–7.66 (m, 4H), 7.25–7.18 (m, 4H).

3. This compound was prepared from 4-bromophenylmagnesium iodide and diethylphosphite according to the procedure for the synthesis of **1**. Column chromatography on silica gel using petroleum ether/ethyl acetate = 1/1 as eluent gave the product as a white solid in 55 % yield. ^{1}H NMR (300 MHz, $CDCl_3$, δ) 8.02 (d, J = 487 Hz, 1H), 7.65 (dd, J = 8.3, 2.2 Hz, 4H), 7.53 (dd, J = 13.2, 8.3 Hz, 4H).

4. To a stirred solution of 4-bromophenol (12.24 g, 70.8 mmol) in 50 mL of dry CH_2Cl_2 at 0 °C was added triethylamine (12.30 mL, 84.9 mmol) and then a solution of tert-butyldimethylsilyl chloride (12.8 g, 84.9 mmol) in 75 mL dry CH_2Cl_2. After that, the reaction mixture was warmed to room temperature, stirred for 24 h, and washed with hydrochloric acid (1 mol/L), followed by washing with brine. The organic phase was dried over Na_2SO_4. The solution was concentrated and the crude product was purified by silica gel column, giving a colorless oil in 99 % yield. ^{1}HNMR (300 MHz, $CDCl_3$, δ): 7.33 (d, J = 8.8 Hz, 2H), 6.73 (d, J = 8.8 Hz, 2H), 0.99 (s, 18H), 0.20 (s, 12H).

5. This compound was prepared from 4-((*tert*-butyldimethylsilyl)oxyl) phenylmagnesium bromide and diethylphosphite according to the procedure for the synthesis of **1**. Column chromatography on silica gel using petroleum ether/ethyl acetate = 1/1 as eluent gave the product as a light yellow oil in 63 % yield. ^{1}HNMR (300 MHz, $CDCl_3$, δ): 8.01 (d, J = 487 Hz, 1H), 7.56 (dd, J = 13.2, 8.4 Hz, 4H), 6.93 (dd, J = 8.4, 2.0 Hz, 4H), 0.98 (s, 18H), 0.22 (s, 12H).

6. A mixture of *p*-dibromobenzene (6.41 g, 27.2 mmol), 3,6-di-tert-butyl-9H-carbazole (2.53 g, 9.1 mmol), CuI (0.17 g, 0.9 mmol), 18-crown-6 (0.09 g, 0.3 mmol), K_2CO_3 (2.50 g, 18.1 mmol) and *o*-dichlorobenzene (2 mL) was heated at 180 °C for 8 h under argon. After cooling to room temperature, the mixture was extracted with CH_2Cl_2, washed with ammonia and then water, and dried over Na_2SO_4. After the solvent had been completely removed, the residue was purified by column chromatography on silica gel using petroleum ether as eluent to give the title product in 80 % yield. ^{1}HNMR (300 MHz, $CDCl_3$, δ): 8.13 (d, J = 1.8 Hz, 2H),

7.70 (d, J = 8.7 Hz, 2H), 7.46 (dd, J = 8.7, 2.1 Hz, 2H) 7.43 (d, J = 8.7 Hz, 2H), 7.31 (d, J = 8.7 Hz, 2H), 1.46 (s, 18H).

7. This compound was prepared from diphenylamine and *p*-diiodobenzene according to the procedure for the synthesis of **6**. Column chromatography on silica gel using petroleum ether as eluent gave the product in 45 % yield. ^{1}HNMR (300 MHz, $CDCl_3$, δ) 7.49 (d, J = 8.5 Hz, 2H), 7.28–7.23 (m, 6H), 7.08–7.01 (m, 6H), 6.82 (d, J = 8.5 Hz, 2H).

8. This compound was prepared from 3,6-di-tert-butyl-9H-carbazole and *p*-diiodobenzene according to the procedure for the synthesis of **6**. Column chromatography on silica gel using petroleum ether as eluent gave the product in 66 % yield. ^{1}HNMR (300 MHz, $CDCl_3$, δ) 8.12 (d, J = 1.8 Hz, 2H), 7.90 (d, J = 8.6 Hz, 2H), 7.46 (dd, J = 8.7, 1.9 Hz, 2H), 7.32 (d, J = 8.7 Hz, 4H), 1.46 (s, 18H).

F-Cz-MON. **2** (0.24 g, 1.0 mmol), **6** (0.48 g, 1.1 mmol), *N*-methylmorpholine (0.15 g, 1.5 mmol) and tetrakis(triphenylphosphine)palladium(0) (0.23 g, 0.2 mmol) was added consecutively to 7 mL toluene under argon. The mixture was then heated at 105 °C for 8 h. After cooling to room temperature, the obtained suspension was directly applied to a silica gel column using petroleum ether/ethyl acetate = 2/1 as eluent to give the crude product. Crystallizing from a mixture of petroleum ether and CH_2Cl_2 gave the pure product as white crystal in a yield of 91 %. ^{1}H NMR (300 MHz, $CDCl_3$, δ): 8.13 (s, 2H), 7.87–7.72 (m, 8H), 7.47 (dd, J = 8.7, 1.7 Hz, 2H), 7.42 (d, J = 8.7 Hz, 2H), 7.23 (d, J = 8.5 Hz, 4H), 1.46 (s, 18H). ^{13}C NMR (75 MHz, $CDCl_3$, δ): 166.5, 163.9, 143.7, 142.1, 138.4, 134.5, 133.5, 129.1, 126.3, 123.8, 116.4, 116.1, 109.1, 34.7, 31.9. ^{31}P NMR ($CDCl_3$, 295 K, δ): 26.80. MALDI-TOF MS: calcd for $C_{38}H_{36}F_2NOP$: 591.25. found: 592.2 $[M + H]^+$. Anal. calcd for $C_{38}H_{36}F_2NOP$: C, 77.14; H, 6.13; N, 2.37. found: C, 77.02; H, 6.01; N, 2.29.

9. This compound was prepared from **5** and **6** according to the procedure for the synthesis of F-Cz-MON. Column chromatography on silica gel using petroleum ether/ethyl acetate = 2/1 as eluent gave the product in 82 % yield. ^{1}H NMR (300 MHz, $CDCl_3$, δ): 8.13 (d, J = 1.7 Hz, 2H), 7.85 (dd, J = 11.4, 8.4 Hz, 2H), 7.68 (dd, J = 8.4, 2.2 Hz, 2H), 7.61 (dd, J = 11.6, 8.6 Hz, 4H), 7.47 (dd, J = 8.6, 1.7 Hz, 2H), 7.42 (d, J = 8.6 Hz, 2H), 6.96 (dd, J = 8.6, 2.3 Hz, 4H), 1.46 (s, 18H), 0.99 (s, 18H), 0.24 (s, 12H).

HO-Cz-MON. **9** was dissolved in a mixture of tetrahydrofuran and hydrochloric acid (2 M) and then stirred at room temperature for 48 h. The mixture was extracted with ethyl acetate and the organic phase was washed with $NaHCO_3$ aqueous and brine and then dried over Na_2SO_4. After complete removal of the solvent, the crude product was purified by column chromatography on silica gel with dichloromethane/methanol = 20/1 as eluent to give the product in quantitative yield. ^{1}H NMR (300 MHz, DMSO-d_6, δ): 10.23 (s, 2H), 8.30 (d, J = 1.4 Hz, 2H), 7.80 (d, J = 7.2 Hz, 4H), 7.52–7.41 (m, 8H), 6.94 (dd, J = 8.6, 2.2 Hz, 4H), 1.41 (s, 18H). ^{13}C NMR (75 MHz, DMSO-d_6) δ: 160.6, 143.0, 140.2, 137.9, 133.6, 133.3, 125.6, 123.8, 123.2, 122.8, 121.7, 116.7, 115.6, 109.2, 34.5, 31.7. ^{31}P NMR (DMSO-d_6, 295 K, δ): 24.97. MALDI-TOF MS: calcd for $C_{38}H_{38}NO_3P$: 587.26. found: 588.3

$[M + H]^+$. Anal. calcd for $C_{38}H_{38}NO_3P$: C, 77.66; H, 6.52; N, 2.38. found: C, 77.50; H, 6.48; N, 2.31.

F-DPA-MON. This compound was prepared from **2** and **7** according to the procedure for the synthesis of F-Cz-MON. Column chromatography on silica gel using petroleum ether/ethyl acetate = 3/2 as eluent gave the product in 82 % yield. ^1H NMR (300 MHz, $CDCl_3$) δ 7.72–7.64 (m, 4H), 7.41–7.28 (m, 6H), 7.19–7.10 (m, 10H), 7.03–6.90 (m, 2H). ^{13}C NMR (75 MHz, $CDCl_3$) δ: 166.3, 163.8, 151.4, 146.4, 134.6, 134.5, 134.4, 133.1, 133.0, 129.6, 128.3, 125.9, 124.6, 122.9, 121.8, 120.0, 119.9, 116.0, 115.9, 115.8, 115.7. ^{31}P NMR ($CDCl_3$, 295 K, δ): 27.25. MALDI-TOF MS: calcd for $C_{30}H_{22}F_2NOP$: 481.1 found: 482.1 $[M + H]^+$. Anal. calcd for $C_{30}H_{22}F_2NOP$: C, 74.84; H, 4.61; N, 2.91. found: C, 74.64; H, 4.55; N, 2.83.

10. This compound was prepared from **1** and **7** according to the procedure for the synthesis of F-Cz-MON. Column chromatography on silica gel using petroleum ether/ethyl acetate = 1/1 as eluent gave the product in 90 % yield. The crude product was used directly in the following step.

HO-DPA-MON. **10** (1.80 g, 3.6 mmol) was dissolved in 20 mL dry dichloromethane. Then the solution was cooled to −78 °C and then BBr_3 (1.8 M in dichloromethane, 8 mL, 14.2 mmol) was added. After that, the reaction mixture was warmed to room temperature, stirred for 24 h, and poured into cold water. The mixture was then extracted with ethyl acetate and the organic layer was washed with $NaHCO_3$ aqueous and water. After drying, the organic layers was concentrated and applied to a silica gel column using CH_2Cl_2/CH_3OH = 20/1 as eluent to give the product with a yield of 80 %. ^1H NMR (300 MHz, DMSO-d_6, δ) 10.12 (s, 2H), 7.41–7.33 (m, 10H), 7.16–7.10 (m, 6H), 6.95–6.90 (m, 2H), 6.89–6.82 (m, 4H). ^{13}C NMR (75 MHz, DMSO-d_6) δ: 160.5, 150.2, 146.2, 133.5, 133.4, 132.8, 132.7, 129.9, 125.7, 124.6, 123.4, 122.3, 119.6, 119.5, 115.6, 115.5. ^{31}P NMR (DMSO-d_6, 295 K, δ): 25.58. MALDI-TOF MS: calcd for $C_{30}H_{24}NO_3P$: 477.1. found: 478.1 $[M + H]^+$. Anal. calcd for $C_{30}H_{24}NO_3P$: C, 75.46; H, 5.07; N, 2.93; found: C, 75.27; H, 5.00; N, 2.88.

Br-Cz-MON. This compound was prepared from **3** and **8** according to the procedure for the synthesis of F-Cz-MON. Column chromatography on silica gel using petroleum ether/ethyl acetate = 2/1 as eluent gave the crude product, which was crystallized in a mixture of petroleum ether and CH_2Cl_2 to give the pure product as white crystal in an overall yield of 78 %. ^1H NMR (300 MHz, $CDCl_3$) δ 8.13 (d, J = 1.8 Hz, 2H), 7.83 (dd, J = 11.5, 8.3 Hz, 2H), 7.74–7.67 (m, 6H), 7.62 (dd, J = 11.2, 8.5 Hz, 4H), 7.46 (dd, J = 8.7, 1.8 Hz 2H), 7.42 (d, J = 8.7 Hz, 2H), 1.46 (s, 18H). ^{13}C NMR (75 MHz, $CDCl_3$) δ: 143.8, 142.3, 138.4, 133.6, 132.2, 132.1, 131.5, 130.5, 127.7, 126.2, 126.1, 123.9, 116.4, 109.1, 34.7, 31.9. ^{31}P NMR ($CDCl_3$, 295 K, δ): 27.14. MALDI-TOF MS: calcd for $C_{38}H_{36}Br_2NOP$: 713.1. found: 714.1 $[M + H]^+$. Anal. calcd for $C_{38}H_{36}Br_2NOP$: C, 63.97; H, 5.09; N, 1.96; found: C, 63.79; H, 4.98; N, 1.90.

Br-DPA-MON. This compound was prepared from **3** and **7** according to the procedure for the synthesis of F-Cz-MON. Column chromatography on silica gel using petroleum ether/ethyl acetate = 3/2 as eluent gave the product in 65 % yield. ^{1}H NMR (300 MHz, $CDCl_3$) δ7.61 (dd, J = 11.1, 2.7 Hz, 4H) 7.54 (dd, J = 11.4, 8.4 Hz, 4H), 7.37 (dd, J = 11.7, 8.7 Hz, 2H), 7.33–7.28 (m, 4H), 7.15–7.10 (m, 6H), 7.01 (dd, J = 8.8, 2.4 Hz, 2H). ^{13}C NMR (75 MHz, $CDCl_3$) 151.5, 146.3, 133.5, 133.4, 132.2, 131.9, 131.8, 131.2, 129.6, 127.2, 125.9, 124.7, 119.9, 119.8. δ: ^{31}P NMR ($CDCl_3$, 295 K, δ): 27.51. MALDI-TOF MS: calcd for $C_{30}H_{22}Br_2NOP$: 601.0. found: 602.0 $[M + H]^+$. Anal. calcd for $C_{30}H_{22}Br_2NOP$: C, 59.73; H, 3.68; N, 2.32; found: C, 59.60; H, 3.58; N, 2.22.

MC. This compound was prepared from **1** and **6** according to the procedure for the synthesis of F-Cz-MON. Column chromatography on silica gel using petroleum ether/ethyl acetate = 2/1 as eluent gave the product in 90 % yield. ^{1}H NMR (300 MHz, $CDCl_3$, δ): 8.13 (d, J = 1.6 Hz, 2H), 7.85 (dd, J = 11.5, 8.5 Hz, 2H), 7.71–7.64 (m, 6H), 7.47 (dd, J = 8.7, 1.8 Hz, 2H), 7.42 (d, J = 8.6 Hz, 2H), 7.03 (dd, J = 8.8, 2.2 Hz, 4H), 3.88 (s, 6H), 1.46 (s, 18H). ^{13}C NMR (75 MHz, $CDCl_3$, δ): 162.5, 143.5, 141.5, 138.6, 134.0, 133.6, 125.9, 124.4, 123.8, 123.3, 116.3, 114.2, 109.2, 55.4, 34.7, 31.2. ^{31}P NMR ($CDCl_3$, 295 K, δ): 28.11. MALDI-TOF MS: calcd for $C_{40}H_{42}NO_3P$: 615.3. found: 616.2 $[M + H]^+$. Anal. calcd for $C_{40}H_{42}NO_3P$: C, 78.02; H, 6.88; N, 2.27. found: C, 77.90; H, 6.79; N, 2.20.

PCzPO. Under argon, F-Cz-MON (0.2959 g, 0.5 mmol), HO-Cz-MON (0.2351 g, 0.5 mmol), K_2CO_3 (0.21 g, 1.5 mmol), *N*, *N*-dimethylacetamide (2.0 mL) and toluene (3.0 mL) were added to a 10 mL flask equipped with a magnetic stir bar, oil-water separator, and gas adapter. The reaction was stirred at 140 °C for 3 h and then 165 °C for 24 h. After cooling to room temperature, the mixture was extracted with CH_2Cl_2. The organic phase was washed with water and then dried over Na_2SO_4. The solution was concentrated and then precipitated in methanol to afford PCzPO as white fiber (0.42 g) in 74 % yield. ^{1}H NMR (400 MHz, $CDCl_3$, δ): 8.12 (s, 2H), 7.91–7.86 (m, 2H), 7.81–7.71 (m, 6H), 7.45–7.40 (m, 4H), 7.20 (d, J = 7.5 Hz, 4H), 1.44 (s, 18H). ^{31}P NMR ($CDCl_3$, 295 K, δ): 27.20. Anal. Calcd for $(C_{38}H_{36}NO_2P)_n$: C, 80.12; H, 6.37; N, 2.46. found: C, 79.96; H, 6.29; N, 2.40.

PDPAPO. This polymer was prepared from F-DPA-MON (0.1926 g, 0.40 mmol) and HO-DPA-MON (0.1910 g, 0.40 mmol) according to the procedure for the synthesis of PCzPO in 55 % yield. ^{1}H NMR (300 MHz, $CDCl_3$, δ): 7.71–7.65 (m, 4H), 7.44–7.38 (m, 2H), 7.31–7.24 (m, 4H), 7.14–7.06 (m, 10H), 7.01 (d, J = 7.2 Hz, 2H). ^{31}P NMR ($CDCl_3$, 295 K, δ):.27.66, Anal. Calcd for $(C_{30}H_{22}NO_2P)_n$: C, 78.42; H, 4.83; N, 3.05; found: C, 78.21; H, 4.73; N, 2.98.

PCzP. To a mixture of $Ni(COD)_2$ (0.55 g, 2 mmol) and 2,2′-bipyridyl (0.31 g, 2 mmol), 0.25 mL 1,5-cyclooctadiene (COD) and 3 mL anhydrous DMF were added under argon. The resulting mixture was kept stirring at 80 °C for 0.5 h. Then a solution of Br-Cz-MON (0.7135 g, 1 mmol) in 3 mL DMF was slowly added to the mixture. The reaction was maintained at 80 °C for 24 h in dark. After cooling to room temperature, the obtained mixture was poured to DCM, then washed intensively with saturated ethylenediaminetetraacetic acid tetrasodium aqueous solution and distilled water. After dried with anhydrous sodium sulfate, the organic phase was concentrated to an appropriate volume and then precipitated in methanol to get a white fiber (0.28 g, yield 50 %). ^{1}H NMR (300 MHz, $CDCl_3$) δ 8.12 (br, 2H), 7.96–7.89 (m, 6H), 7.82–7.73 (m, 6H), 7.47–7.43 (m, 4H), 1.45 (br, 18H). ^{31}P NMR ($CDCl_3$, 295 K, δ): 27.58. Anal. Calcd for $(C_{38}H_{36}NOP)_n$: C, 82.43; H, 6.55; N, 2.53; found: C, 82.31; H, 6.50; N, 2.43.

PDPAP. This polymer was prepared from Br-DPA-MON (0.8000 g, 1.33 mmol) according to the procedure for PCzP in 31 % yield. ^{1}H NMR (300 MHz, $CDCl_3$) δ 7.80 (dd, $J = 10.2$, 8.4 Hz, 1H), 7.67 (d, $J = 6.9$ Hz, 4H), 7.44 (dd, $J = 10.8$, 9.0 Hz, 2H), 7.31–7.27 (m, 4H), 7.13–7.06 (m, 6H), 7.01 (d, $J = 7.3$ Hz, 2H). ^{31}P NMR ($CDCl_3$, 295 K, δ): 28.04. Anal. Calcd for $(C_{30}H_{22}NOP)_n$: C, 81.25; H, 5.00; N, 3.16; found: C, 81.07; H, 4.92; N, 3.06.

References

1. Baldo MA, O'Brien DF, You Y et al (1998) Highly efficient phosphorescent emission from organic electroluminescent devices. Nature 395:151–154
2. Kawamura Y, Yanagida S, Forrest SR (2002) Energy transfer in polymer electrophosphorescent light emitting devices with single and multiple doped luminescent layers. J Appl Phys 92:87–93
3. Gong X, Ma WL, Ostrowski JC et al (2004) White electrophosphorescence from semiconducting polymer blends. Adv Mater 16:615–619
4. Sudhakar M, Djurovich PI, Hogen-Esch TE et al (2003) Phosphorescence quenching by conjugated polymers. J Am Chem Soc 125:7796–7797
5. Chou HH, Cheng CH (2010) A highly efficient universal bipolar host for blue, green, and red phosphorescent OLEDs. Adv Mater 22:2468–2471
6. Polikarpov E, Swensen JS, Chopra N et al (2009) An ambipolar phosphine oxide-based host for high power efficiency blue phosphorescent organic light emitting devices. Appl Phys Lett 94:223304
7. Jeon SO, Yook KS, Joo CW et al (2010) High-efficiency deep-blue-phosphorescent organic light-emitting diodes using a phosphine oxide and a phosphine sulfide high-triplet-energy host material with bipolar charge-transport properties. Adv Mater 22:1872–1876
8. van Dijken A, Bastiaansen J, Kiggen NMM et al (2004) Carbazole compounds as host materials for triplet emitters in organic light-emitting diodes: polymer hosts for high-efficiency light-emitting diodes. J Am Chem Soc 126:7718–7727
9. Brunner K, van Dijken A, Borner H et al (2004) Carbazole compounds as host materials for triplet emitters in organic light-emitting diodes: tuning the HOMO level without influencing the triplet energy in small molecules. J Am Chem Soc 126:6035–6042
10. Chen YC, Huang GS, Hsiao CC et al (2006) High triplet energy polymer as host for electrophosphorescence with high efficiency. J Am Chem Soc 128:8549–8558
11. Zhang K, Tao Y, Yang C et al (2008) Synthesis and properties of carbazole main chain copolymers with oxadiazole pendant toward bipolar polymer host: tuning the HOMO/LUMO level and triplet energy. Chem Mater 20:7324–7331
12. Mathai MK, Choong V-E, Choulis SA et al (2006) Highly efficient solution processed blue organic electrophosphorescence with 14 Lm/W luminous efficacy. Appl Phys Lett 88:243512
13. Shao S, Ding J, Ye T et al (2011) A novel, bipolar polymeric host for highly efficient blue electrophosphorescence: a non-conjugated poly(aryl ether) containing triphenylphosphine oxide units in the electron-transporting main chain and carbazole units in hole-transporting side chains. Adv Mater 23:3570–3574
14. Nielsen KT, Bechgaard K, Krebs FC (2005) Removal of palladium nanoparticles from polymer materials. Macromolecules 38:658–659

15. Sapochak LS, Padmaperuma AB, Cai XY et al (2008) Inductive effects of diphenylphosphoryl moieties on carbazole host materials: design rules for blue electrophosphorescent organic light-emitting devices. J Phys Chem C 112:7989–7996
16. Ding J, Wang Q, Zhao L et al (2010) Design of star-shaped molecular architectures based on carbazole and phosphine oxide moieties: towards amorphous bipolar hosts with high triplet energy for efficient blue electrophosphorescent devices. J Mater Chem 20:8126–8133
17. Yeh HC, Chien CH, Shih PI et al (2008) Polymers derived from 3,6-fluorene and tetraphenylsilane derivatives: solution-processable host materials for green phosphorescent OLEDs. Macromolecules 41:3801–3807

Chapter 3
Blue/Yellow Electrophosphorescent Polymers Based on Polyarylether Hosts

3.1 Background

Although high-efficiency PhPLEDs have been realized by doping phosphorescent dyes into polymer hosts as presented in Chap. 2, these physical blend systems suffer from phase segregation under long-term device operation. To solve this problem, developing electrophosphorescent polymers (PhPs), where the phosphorescent dyes are incorporated into the polymeric main chain or side chain via covalent bonds, is necessary (Fig. 3.1). In literatures [1–8], most efforts have been paid to the development of red and green PhPs, and their luminous efficiencies have reached up to 26.7 cd A^{-1} [5] and 33.9 cd A^{-1} [8], respectively. In contrast, the efficiency of blue and yellow PhPs are low, which have been kept at 5.5 cd A^{-1} [9, 10] and 6. 1 cd A^{-1} [11], respectively. These efficiencies are far from practical application. As one of the three primary colors, blue PhPs are indispensable components for low-cost full-color displays as well as white light sources. Similarly, yellow emitters are also highly desirable because they can be used not only to generate white electroluminescence (EL) when mixed with sky-blue dyes, but also to improve the power efficiency of white OLEDs (WOLEDs) [12, 13]. In view of these two points, the de nova design of blue and yellow PhPs is highly desirable.

As discussed in Chap. 1, the low efficiency of blue and yellow PhPs is mainly caused by the lack of suitable polymer hosts which can be used as the molecular design platform. In theory, some critical criteria have to be taken into account for the platform. Of primary importance is that the triplet energy of the polymer backbone should be higher than the phosphorescent dyes (>2.70 eV for blue ones) [14], so that triplet excitons can be effectively confined on the phosphors. Simultaneously, the HOMO and LUMO levels should match well with the Fermi levels of the electrodes to facilitate charge injection. However, it is difficult to meet these two requirements at the same time, because polymers with high triplet energy usually suffer from wide bandgap and thus charge injection problems. For example, poly(*N*-vinylcarbazole) (PVK) with a high triplet energy of 3.0 eV is often used as the

S. Shao, *Electrophosphorescent Polymers Based on Polyarylether Hosts*,
Springer Theses, DOI 10.1007/978-3-662-44376-7_3

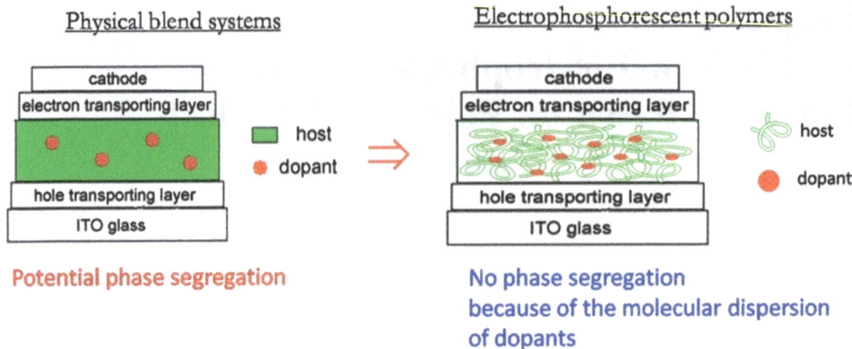

Fig. 3.1 Comparison of the physical blend systems and the electrophosphorescent polymers

backbone for blue PhPs. Regretfully, it has a deep HOMO level of −5.9 eV and a shallow LUMO level of −2.1 eV, which leads to large charge injection barriers and poor device performance (5.5 cd A^{-1}) [9].

3.2 Molecular Designs

In Chap. 2, we have realized a tradeoff between HOMO/LUMO levels and triplet energy on a novel, polyarylether host PCzPO by adopting a poly(arylene ether phosphine oxide) scaffold. Compared with PVK, its HOMO and LUMO levels are tuned to −5.7 eV and −2.3 eV, respectively, to improve the charge injection capability. In the meantime, the triplet energy of PCzPO remains as high as 2.96 eV. With PCzPO as the host of a typical blue phosphor FIrpic, a promising luminous efficiency of 23.3 cd A^{-1} is attained. Encouraged by this preliminary result, we attempt to introduce FIrpic and a 2-(fluoren-2-yl)-1*H*-benzoimidazole-based Ir complex ($(fbi)_2Ir(acac)$) as the blue and yellow dopant, respectively, to the side chain of PCzPO to construct blue and yellow PhPs (Fig. 3.2) [15, 16].

Fig. 3.2 Molecular designs for the *blue/yellow* PhPs

3.3 Results and Discussions

3.3.1 Blue Electrophosphorescent Polyarylethers

Initial attempts Our initial attempt involved directly introducing 5 mol% FIrpic to the side chain of PCzPO to construct the blue polymer PB'-05. As shown in Scheme 3.1, we first synthesized the hydroxyl-containing blue complex FIrpicOH, which was followed by reaction with **3** to give the monomer Blue'-MON carrying the blue complex. Then we expect that polymerization of Blue'-MON, F-Cz-MON, and HO-Cz-MON under nucleophilic substitution polycondensation condition can afford the target PhPs. Unfortunately, the resultant polymer PB'-0.05 is not the desired blue PhP.

Reagents and conditions:

i) *p*-bromoanisole, Pd(PPh3)4, *N*-methylmorpholine and toluene, 105 °C; ii) BBr_3, CH_2Cl_2, −78 oC; iii) K_2CO_3, 1,8-dibromooctane, acetone, reflux; iv) K_2CO_3, FIrpicOH, DMF, 60 °C.

As shown in Fig. 3.3, the PL spectrum of PB'-0.05 exhibits a red-shift of 12 nm in film state relative to the physical blend that is composed of 5 mol% FIrpic mixed with PCzPO. Similarly, the EL spectrum of PB'-0.05 is also red-shifted by about

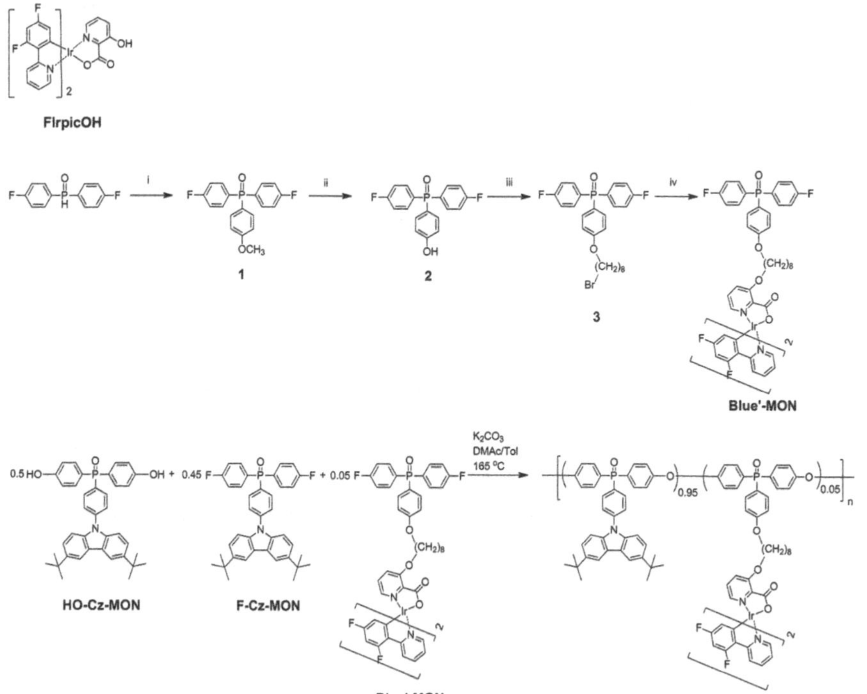

Scheme 3.1 Synthetic routes of PB'-0.05

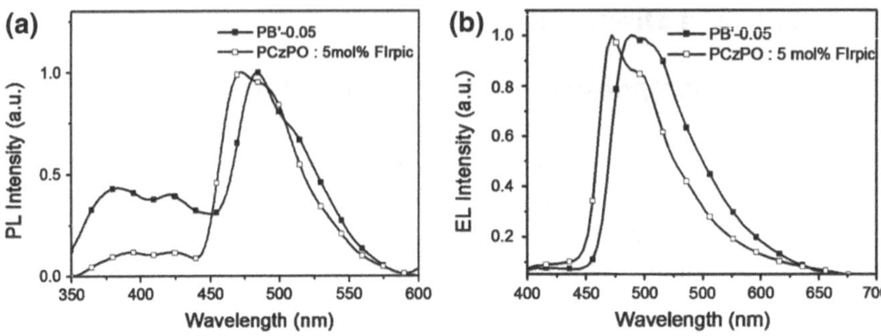

Fig. 3.3 Comparison of the PL (**a**) and EL (**b**) spectra of PB'−0.05 and the physical blend of FIrpic and PCzPO

20 nm compared with the corresponding physical blends. The reason of this phenomenon is not very clear yet. We only speculate that the F atoms of FIrpic may be not tolerant to the harsh polycondensation condition involving the adoption of a base in a polar aprotic solvent at a high reaction temperature of 165 °C for 24 h.

Fluorinated poly(arylene ether phosphine oxide) backbone To solve this problem, we attempt to enhance the reaction activity of the monomers by introducing additional fluorine atoms into the *ortho*-position of the fluorine groups of the F-Cz-MON monomer. According to the nucleophilic substitution mechanism, four adjacent electron-withdrawing fluorine atoms can substantially enhance the activity of the C–F bonds *para* to P atom. With this change, the resultant fluorinated poly (arylene ether phosphine oxide) FPCzPO were successfully prepared under a moderate temperature of 120 °C (Scheme 3.2). Consequently, FIrpic is compatible with this milder condition, and undesired color alteration of blue PhPs is avoided.

Reagents and conditions:

v) (1) Mg, tetrahydrofuran, (2) diethylphosphite, (3) NH_4Cl; vi) 9-(4-bromophenyl)-3,6-di-tert-butyl- 9H-carbazole, $Pd(PPh_3)_4$, N-methylmorpholine and toluene, 100 °C;

Here several issues should be noted:

(1) Although six C–F bonds exist, the nucleophilic reaction occurs merely on the two activated C–F bonds *para* to P atom, which makes sure that only soluble linearly polymers are obtained. This speculation is supported by the results reported before [17], as well as the ^{13}C NMR and ^{19}F NMR spectra of FPCzPO. In the ^{13}C NMR spectrum (Fig. 3.4a), only one set of dd peaks (δ 156.06 (dd, $^1J_{C\text{-}F}$ = 259.4 Hz, $^3J_{C\text{-}P}$ 18.7 Hz)) is observed for the C–F atoms, indicating that the polymerization occurs only at the fluorine atoms *para* to the P = O bond. Similarly, in the ^{19}F NMR spectrum, only one strong singlet peak appears at −8.02 ppm (vs. fluorobenzene) (Fig. 8.4b). If the C–F bonds *meta* to P atom take part in the polymerization, there would be at least two sets of signals.

(2) The incorporation of additional F atoms does not obviously affect the original electronic properties of the main chain. As presented in Fig. 3.5, both FPCzPO and PCzPO show similar absorption peaks at 296/332/342 nm in dichloromethane. The

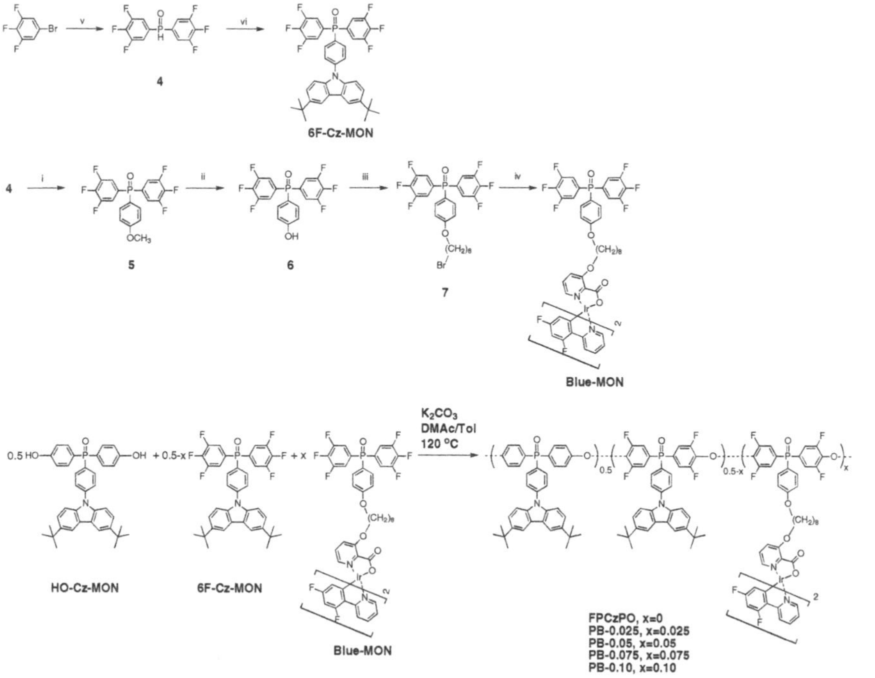

Scheme 3.2 Synthetic routes of FPCzPO and PB-0.025 ~ PB-0.10

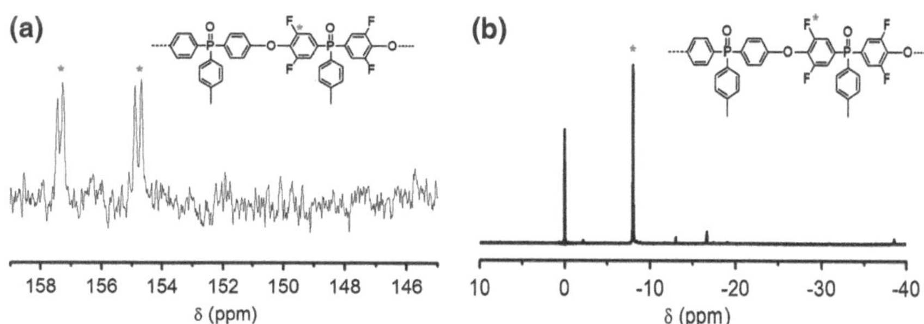

Fig. 3.4 ^{13}C NMR (**a**) and ^{19}F NMR (**b**) spectra of FPCzPO

PL peak of FPCzPO in toluene is located at 406 nm, which is only slightly red-shifted compared with PCzPO. In addition, the FPCzPO film shows two emission peaks at 397/417 nm, which are again close to those of PCzPO. Furthermore, the oxidation and reduction potential of FPCzPO are 0.9 eV and −2.5 eV, respectively (Fig. 3.6). Accordingly, the HOMO and LUMO levels are estimated to be −5.7 eV and −2.3 eV, respectively, which are very close to the values of PCzPO. Most importantly, FPCzPO and PCzPO show almost the same phosphorescent spectra at 77 K,

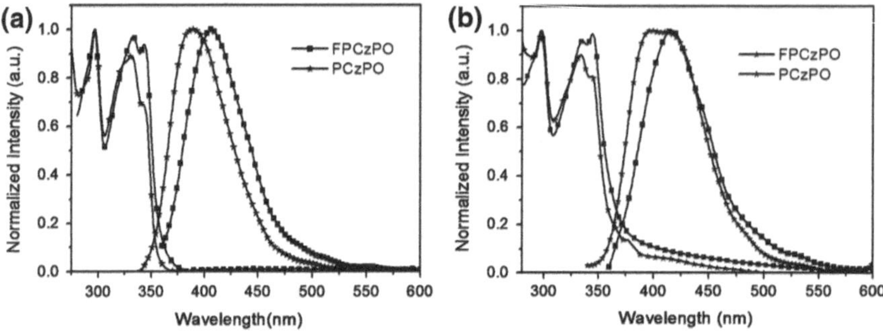

Fig. 3.5 **a** Absorption (in dichloromethane, 10^{-5} mol L^{-1}) and PL spectra (in toluene, 10^{-5} mol L^{-1}) of FPCzPO and PCzPO; **b** Absorption and PL spectra of FPCzPO and PCzPO in film state

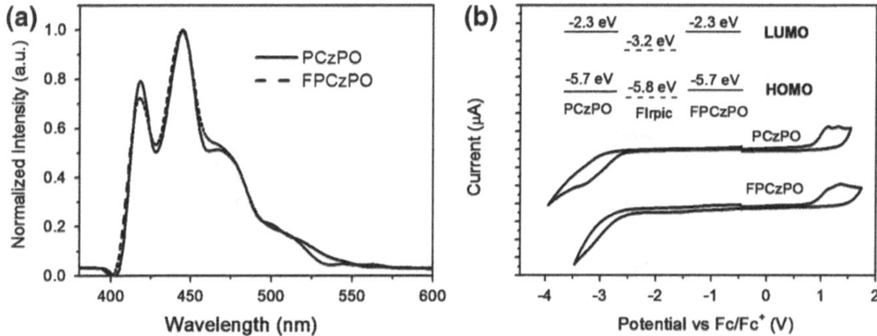

Fig. 3.6 Phosphorescence spectra (77 K) (**a**) and cyclic voltammetry curves of FPCzPO and PCzPO (**b**)

indicating that they exhibit nearly the same E_{T}s ($\sim$2.96 eV). In view of these results, we speculate that FPCzPO can act as an excellent host material as PCzPO does.

Synthesis and Characterization of Blue PhPs To continue the effort of constructing blue PhPs, FIrpic was attached to the sidechain of FPCzPO. As shown in Scheme 3.2, a hexafluorinated comonomer Blue-MON carrying blue complex FIrpic is synthesized according to the procedure similar as Blue'-MON. Based on this monomer as well as 6F-Cz-MON and HO-Cz-MON, a series of blue PhPs (PB-0.025 $\sim$ PB-0.10) have been synthesized under the above-mentioned mild polymerization condition. Ir-loadings of the polymers are tailored by the feed ratio between 6F-Cz-MON and Blue-MON. Work-up of simply washing by water, and then precipitating in methanol affords the polymers pure enough to be used in PLEDs without any residual catalyst contamination, which is always a big trouble in transition-metal-catalyzed conjugated polymers [18]. These polymers possess high solubility in common organic solvents, such as dichloromethane, chloroform, and chlorobenzene. All the polymers show high decomposition temperatures (T_{d}) above 370 °C. Meanwhile, the actual content of FIrpic in the polymers is estimated

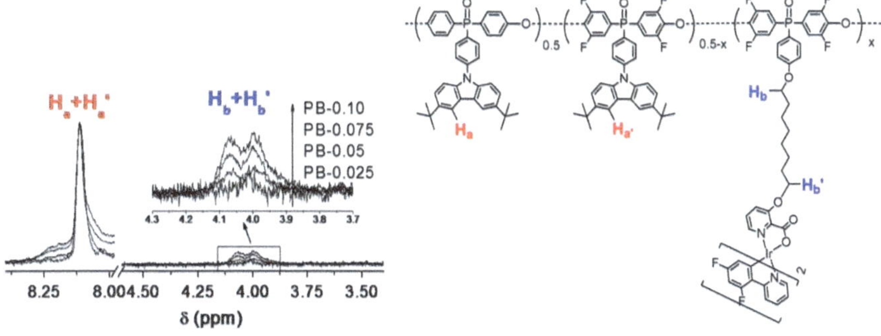

Fig. 3.7 ^{1}H NMR of the *blue* PhPs

Table 3.1 Characterization of FPCzPO and the *blue* PhPs

Polymer	M_n	PDI	T_d	Ir-complex loading (mol%)	
				Feed ratio	Calculated
FPCzPO	7,800	1.64	493	–	–
PB-0.025	7,500	1.73	393	0.025	0.018
PB-0.05	7,700	1.70	382	0.050	0.042
PB-0.075	7,600	1.68	383	0.075	0.070
PB-0.10	7,800	1.67	379	0.10	0.090

by comparing the integration of the ^{1}H NMR signals at 3.95–4.10 ppm ($-CH_2O-$ groups in the alkyl spacer) with the singlet peak at 8.10 ppm related to the protons of carbazole units (Fig. 3.7). As listed in Table 3.1, the Ir loading is found to be slightly lower than the feed ratio.

Photophysical properties Absorption spectra of the PhPs in dichloromethane and their PL spectra in toluene and solid states are shown in Fig. 3.8. All the PhPs display strong π-π* absorption peaks at 297/329/344 nm, similar to those of FPCzPO. The low energy absorption bands at 375–450 nm of the metal-to-ligand charge transfer (MLCT) transition of FIrpic are too weak to be distinguished due to its small absorption coefficient and the low Ir loading. PL spectra of the PhPs show two distinct PL peaks at about 402 and 472 nm, which are assigned to the emissions of FPCzPO and FIrpic, respectively. We note that, unlike PB′-0.05, PB-0.025 ~ PB-0.10 shows normal phosphorescence from FIrpic, indicating that FIrpic is successfully introduced into the FPCzPO side chain without destroying the structure. Furthermore, when the Ir loading increases, the relative emission intensity of FIrpic increases obviously. This demonstrates that intramolecular Förster energy transfer occurs from FPCzPO to FIrpic, which is reasonable because the absorption of FIrpic overlaps well with the PL spectrum of FPCzPO (Fig. 3.9) [19].

Similar as in solution, the PhPs show nearly the same absorption spectra in film. However, for their PL spectra, the emission of FIrpic becomes dominant on going

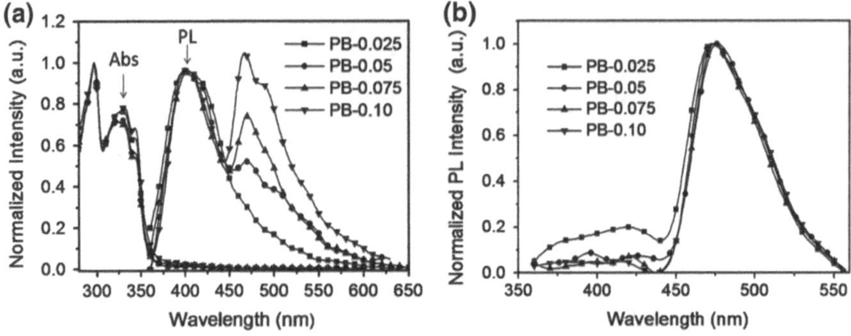

Fig. 3.8 Absorption (in DCM) and PL spectra (in toluene) of PB-0.025 ~ PB-0.10 (**a**), and their PL spectra in films (**b**). Reprinted with permission from Ref. [15]. Copyright 2012 American Chemical Society

from solution to solid state, suggesting that both intra- and intermolecular Förster energy transfer contribute to the emission. Noticeably, even in PB-0.025, the emission of FPCzPO is quenched to a large extent. This observation is different from that in PCzSiIr2.5 [10], where the emission of the polymeric host is much more intense despite the same Ir complex content. This difference can be explained by the higher triplet energy of FPCzPO than P36HCTPSi, which prevents the loss of triplet excitons caused by back transfer (from the Ir complex to the polymer mainchain) process.

Electroluminescent properties To explore the EL performance, PB-0.025 ~ PB-0.10 are used as the emitting layer (EML) to fabricate PhPLEDs with a configuration of ITO/PEDOT:PSS (40 nm)/EML (40 nm)/TPCz (50 nm)/LiF (1 nm)/Al (100 nm). Figure 3.10 shows the EL spectra of the PhPs. They all reveal similar EL profiles with CIE coordinates of (0.18, 0.33). Moreover, the EL spectra are very close to their PL counterparts, with no emission from FPCzPO or TPCz observed. We speculate that Förster energy transfer rather than charge trapping dominants in

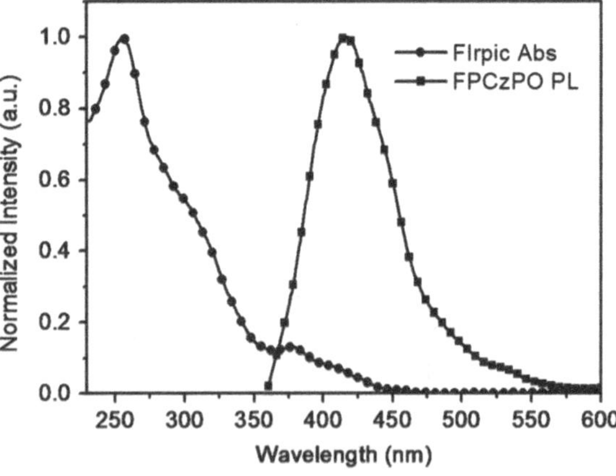

Fig. 3.9 Absorption spectra of FIrpic and PL spectra of FPCzPO

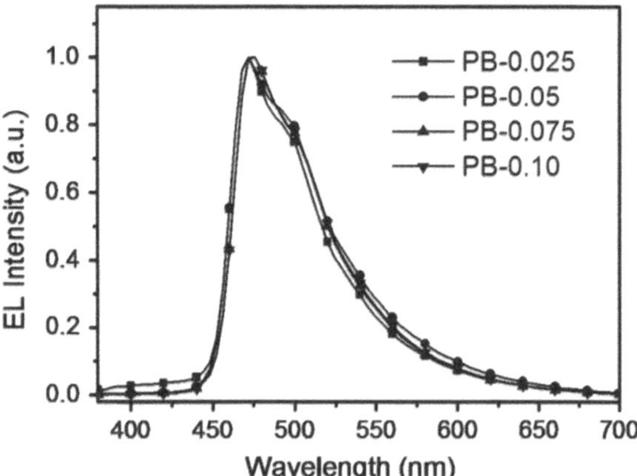

Fig. 3.10 EL spectrum of PB-0.05. Reprinted with permission from Ref. [15]. Copyright 2012 American Chemical Society

the formation of excitons in the EL process. This hypothesis is confirmed by the Ir complex concentration dependence of the *J–V* characteristics. As shown in Fig. 3.11a, except for PB-0.025, the *J–V* curves shift toward a lower driving voltage when the Ir loading increases from 5 mol% to 10 mol%.

As depicted in Fig. 3.11b and Table 3.2, the Ir loading shows a weak influence on the device performance. The luminous efficiency first rises from 17.1 cd A^{-1} of PB-0.025 to 19.4 cd A^{-1} of PB-0.05, and then decreases to 16.5 cd A^{-1} for PB-0.10. A variance of only ~15 % is observed as the Ir loading changes from 2.5 mol % to 10 mol%. Among these PhPs, PB-0.05 gives the best performance with a peak η_l of 19.4 cd A^{-1}, a peak η_p of 10.0 lm W^{-1}, a peak EQE of 9.0 %, and a maximum brightness of 5508 cd m^{-2}. Even at a brightness level of 1000 cd m^{-2}, the η_l still retains as high as 14.5 cd A^{-1} (EQE 6.7 %). To the best of our knowledge, this is the highest efficiency for blue PhPs ever reported, which is comparable with that of the physical blend of PCzPO and FIrpic. More importantly, compared with the blue PhPs using PVK as the main chain [9], an about 3.5-fold improvement of η_l has been realized. Considering that the grafted phosphor is the same in these studies

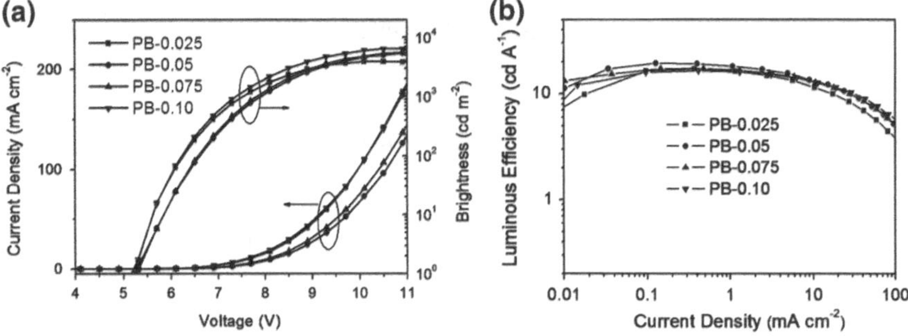

Fig. 3.11 *J-V–B* and η_l *-J* curves of PB-0.025–PB-0.10. Reprinted with permission from Ref. [15]. Copyright 2012 American Chemical Society

(FIrpic), we propose that the state-of-art performance of our polymers is attributed to the use of the fluorinated poly(arylene ether phosphine oxide) backbone.

In conclusion, a series of blue PhPs with fluorinated poly(arylene ether phosphine oxide) as the backbone have been successfully synthesized under a compatible condition with FIrpic enabled by the activation of the monomer by fluorination. These PhPs emit normal EL emission from FIrpic, and show a promising luminous efficiency as high as 19.4 cd A^{-1} for the first time. We believe that, this work will pave the way for the development of high-performance blue PhPs as well as all-phosphorescent single white polymers in the future.

3.3.2 Yellow Electrophosphorescent Polyarylethers

Similar as in blue PhPs, the fluorinated polyarylether is adopted as the backbone for the yellow PhPs. By grafting ($(fbi)_2Ir(acac)$) with different contents to its side chain, we have synthesized a series of novel yellow PhPs (PO-0.01–PO-0.04).

Synthesis and characterization Scheme 3.3 shows the synthetic routes of the yellow PhPs. First, the reaction between 2-(9,9-diethyl-9*H*-fluorenyl-2-yl)-1-phenyl-1*H*-benzo[d]imidazole (**3**) and $IrCl_3 \cdot 3H_2O$ in 2-ethoxyethanol/H_2O afforded the chloro-bridged dimer **4** in a high yield. Then, the auxiliary ligand **10**, which was prepared by coupling aryl iodide **9** with bisarylphosphine oxide **6** under the palladium-catalyzed Hirao reaction, reacted with the dimer **4** to provide the key monomer Y-MON containing the yellow Ir complex, $(fbi)_2Ir(acac)$. Together with the comonomer HO-Cz-MON and 6F-Cz-MON, polymer PO-0.01 ~ PO-0.04 were finally synthesized through a modified nucleophilic aromatic substitution polymerization at 120 °C with K_2CO_3 as a base and DMAc/toluene as a solvent. The content of the incorporated Ir complex into the polymer was controlled via the feed ratio. Work-up of washing by water and then precipitating in methanol afforded the obtained polymers pure enough to be used in OLEDs without any residual catalyst contamination. The number molecular weight (M_n) and the polydispersity index (PDI) of the polymers were measured by gel permeation chromatography (GPC), and the data are listed in Table 3.3. Moderate M_n of 6600–7400 Da for all the yellow PhPs are obtained, and they do not alter obviously with the Ir loading. The polymers exhibit good solubility in common organic

Table 3.2 Device performance of PB-0.025–PB-0.10. Reprinted with permission from Ref. [15]. Copyright 2012 American Chemical Society

Polymer	V_{on} [V]	$\eta_{l,\ max}$ [cd A^{-1}]	$\eta_{p,\ max}$ [lm W^{-1}]	EQE (%)	L_{max} [cd m^{-2}]	CIE (x, y)
PB-0.025	5.1	17.1	9.0	8.0	3927	(0.18, 0.33)
PB-0.05	5.3	19.4	10.0	9.0	5508	
PB-0.075	5.3	17.3	8.9	8.0	5531	
PB-0.10	5.3	16.5	8.8	7.6	6367	

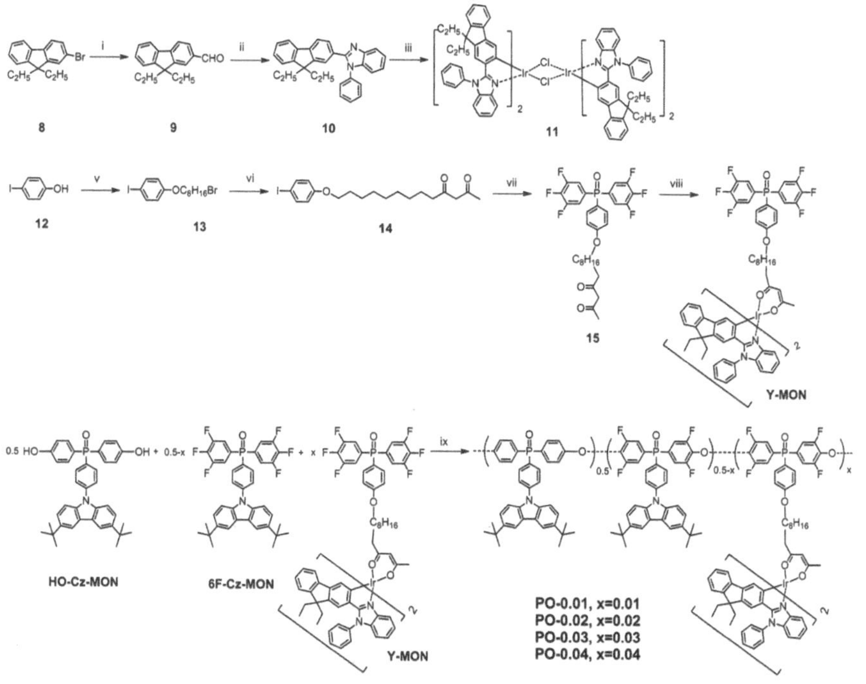

Scheme 3.3 Synthetic routes of the *yellow* PhPs

solvents, such as dichloromethane, tetrahydrofuran, toluene and chlorobenzene, and high quality films can be formed via spin-coating. Furthermore, thermogravimetric analysis (TGA) indicates that they are thermally stable with the decomposition temperatures (T_d) higher than 380 °C. And their glass-transition temperatures (T_g) determined by differential scanning calorimetry (DSC) are higher than 240 °C, implying that the polymers possess excellent morphological stability.

Reagents and conditions:

i. 1) *n*-BuLi, tetrahydrofuran, −78 °C, 2) *N,N*-dimethylformamide; ii. 2-aminodiphenylamine, H_2O_2, concentrated hydrochloric acid, 2-ethoxyethanol, 25 °C; iii. $IrCl_3$-$3H_2O$, 2-ethoxyethanol/H_2O, reflux; iv 1) Mg, tetrahydrofuran,

Table 3.3 Structure characterization data for PO-0.01 ~ PO-0.04. Reproduced from Ref. [16] with permission from The Royal Society of Chemistry

Polymer	M_n	PDI	T_d (°C)	T_g (°C)	Ir-complex loading (mol %)	
					Feed ratio	Calculated
PO-0.01	7, 400	1.63	400	250	0.01	–
PO-0.02	6, 600	1.63	404	248	0.02	–
PO-0.03	6, 900	1.54	401	254	0.03	0.034
PO-0.04	7, 100	1.64	381	241	0.04	0.044

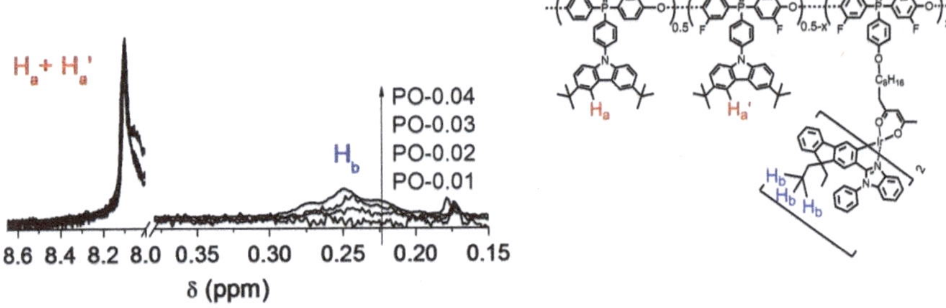

Fig. 3.12 ^{1}H NMR spectra of the *yellow* polymers

2) diethylphosphite, 3) NH_4Cl; v 1,8-dibromooctane, K_2CO_3, acetone, reflux; vi. 1) acetylacetone, NaH, tetrahydrofuran, 2) *n*-BuLi, 0 °C, 3) **8**, tetrahydrofuran; vii **6**, $Pd(PPh_3)_4$, 4-methylmorpholine, toluene, 100 °C; viii **4**Na_2CO_3, 2-ethoxyethanol, reflux; ix K_2CO_3*N*,*N*-dimethylacetamide, toluene, 120 °C

The actual content of $(fbi)_2Ir(acac)$ incorporated into the polymers is determined by their ^{1}H NMR spectra, as demonstrated in the blue PhPs. Figure 3.12 presents two sets of characteristic peaks of the ^{1}H NMR spectra of the polymers. The signals at 8.0–8.2 ppm can be assigned to the 4- and 5-position hydrogens on carbazole units. And the peaks located at 0.20–0.30 ppm, which become discernible when the feed ratio of Ir complex is higher than 2 mol%, are corresponding to the methyl groups of $(fbi)_2Ir(acac)$. By comparing their individual integration, the Ir loading is estimated, and the data are listed in Table 3.3. As can be clearly seen, the calculated content of $(fbi)_2Ir(acac)$ is found to be very close to the feed ratios. This suggests that the added Y-MON monomers are almost fully incorporated into the polymers during this polymerization.

Figure 3.13a presents the absorption spectra of PO-0.01 ~ PO-0.04 in dichloromethane solution. PO-0.01 ~ PO-0.04 show similar strong absorption peaks at 296/331/344 nm from the π-π* transition of the host FPCzPO. Meanwhile, the low

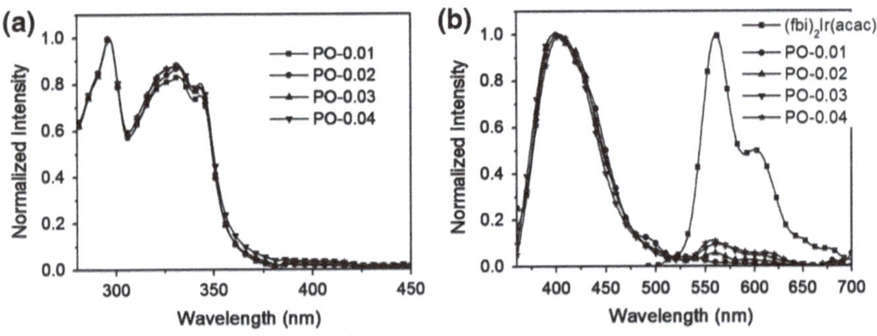

Fig. 3.13 **a** UV-visible absorption spectra of the *yellow* PhPs in dichloromethane (10^{-5} mol L^{-1}); **b** PL spectra of $(fbi)_2Ir(acac)$ and the *yellow* PhPs in toluene solution (10^{-5} mol L^{-1}). Reproduced from Ref. [16] with permission from The Royal Society of Chemistry

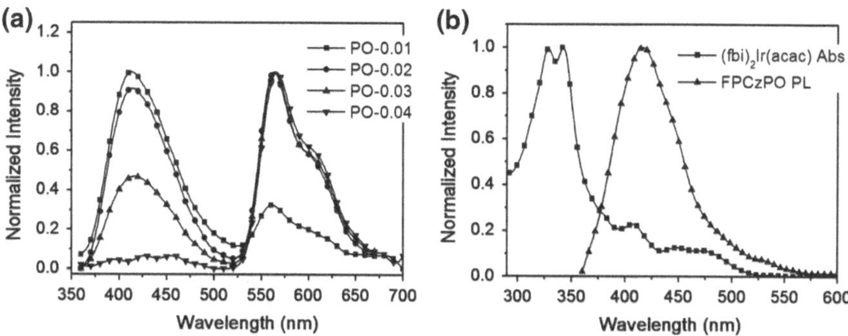

Fig. 3.14 **a** PL spectra of the yellow PhPs in film; **b** Overlap of the absorption spectrum of $(fbi)_2Ir$ (acac) in dichloromethane and PL spectrum of FPCzPO in film. Reproduced from Ref. [16] with permission from The Royal Society of Chemistry

energy absorption band in the range of 375–450 nm is negligible, which is ascribed to the metal-to-ligand charge transfer transition of the Ir complex $(fbi)_2Ir(acac)$. This is understandable when the low Ir loading is taken into account. Based on the same reason, the emissions are mainly from FPCzPO in their PL spectra in toluene solution for PO-0.01 ~ PO-0.04 (Fig. 3.13b). Furthermore, the intensity of the long-wavelength emission from $(fbi)_2Ir(acac)$ increases slightly with the content of the Ir complex. For example, when the Ir loading is as high as 4 mol%, the emission of FPCzPO still dominates the whole spectrum of PO-0.04, indicative of inefficient intramolecular Förster energy transfer in this case.

Compared with those in solution, the PL spectra in film of PO-0.01 ~ PO-0.04 (Fig. 3.14a) show distinct profiles with much higher emission intensity of $(fbi)_2Ir$ (acac). And the relative intensity ratio between $(fbi)_2Ir(acac)$ and FPCzPO enhances greatly when the Ir loading varies. With an Ir complex content up to 4 mol% in PO-0.04, the emission of FPCzPO is almost quenched completely, and only the emission from $(fbi)_2Ir(acac)$ exists. These observations demonstrate that the intermolecular Förster energy transfer from the host to the Ir complex happens efficiently in film. According to the energetic resonance theory [19], the possibility of this Förster energy transfer can be supported by the good overlap between the absorption spectrum of $(fbi)_2Ir(acac)$ and PL spectrum of FPCzPO, as depicted in Fig. 3.14b.

Electroluminescent Performance To investigate their EL performance, PO-0.01 ~ PO-0.04 were used as the emitting layer (EML) to fabricate PhPLEDs with a configuration of glass/ITO/PEDOT:PSS (40 nm)/EML(40 nm)/TPCz (50 nm)/LiF (1 nm)/Al (100 nm) (Fig. 3.15a). As presented in Fig. 3.15b, PO-0.01 ~ PO-0.04 display EL emission maxima at 566 nm typical for the emission of $(fbi)_2Ir(acac)$, and the emission from FPCzPO at 435 nm completely disappears when the Ir loading reaches 2 mol%. Consequently, except for PO-0.01 (CIE: (0.45, 0.41)), PO-0.02, PO-0.03, and PO-0.04 exhibit the same CIE coordinates of (0.53, 0.46). In addition, the EL spectra are different from their PL counterparts (especially for PO-0.02), where distinct FPCzPO emission can be observed. Such a difference

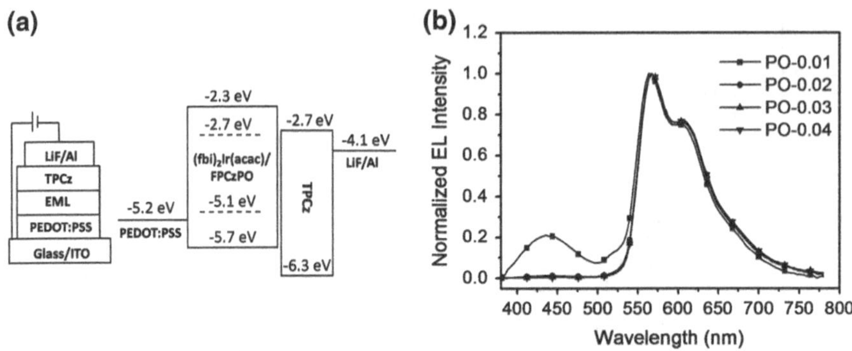

Fig. 3.15 **a** Device structure and the HOMO/LUMO energy levels of $(fbi)_2Ir(acac)$ and FPCzPO; **b** EL spectra of the *yellow* PhPs. Reproduced from Ref. [16] with permission from The Royal Society of Chemistry

between the PL and EL suggests that besides Förster energy transfer, another exciton-forming pathway may exist in the EL process. As shown in Fig. 3.15a, we can see that both the HOMO and LUMO levels of $(fbi)_2Ir(acac)$ (HOMO: −5.1 eV; LUMO: −2.7 eV) fall within those of FPCzPO (HOMO: −5.7 eV; LUMO: −2.3 eV). Under electric excitation, holes and electrons can be directly injected into $(fbi)_2Ir(acac)$ or first injected into FPCzPO and then readily trapped by $(fbi)_2Ir$ (acac). As for both cases, excitons are directly formed on $(fbi)_2Ir(acac)$ to result in efficient triplet emissions. In other words, charge trapping mechanism plays a paramount role during the EL process.

This can be further verified by the driving voltage dependence of the current density. As shown in Fig. 3.16a, the current density at the same driving voltage decreases gradually as the Ir loading increases, indicative of the existence of the above-mentioned charge trapping process. Additionally, the Ir loading also affects the luminous efficiency. Figure 3.16b depicts the luminous efficiency versus current density curves of the devices and Table 3.4 summarizes the corresponding device

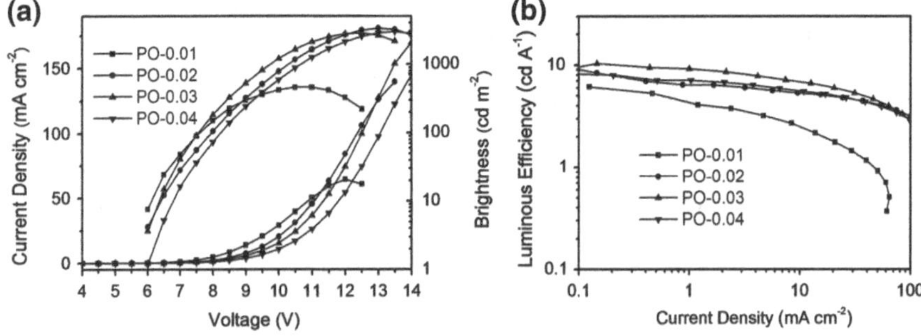

Fig. 3.16 *J-V–B***a** and η_l *-J***b** curves of the devices. Reproduced from Ref. [16] with permission from The Royal Society of Chemistry

Table 3.4 Device performance of PO-0.01 ~ PO-0.04. Reproduced from Ref. [16] with permission from The Royal Society of Chemistry

Polymer	V_{on} [a] [V]	$\eta_{l,\ max}$ [b] [cd A^{-1}]	$\eta_{p,\ max}$ [c] [lm W^{-1}]	EQE_{max} [d] (%)	L_{max} [e] [cd m^{-2}]	CIE [f] (x, y)
PO-0.01	5.5	6.2	3.2	2.4	463	0.45, 0.41
PO-0.02	5.5	8.4	4.1	3.3	3346	0.53, 0.46
PO-0.03	5.5	10.4	5.0	4.1	2842	0.53, 0.46
PO-0.04	6.0	8.3	4.0	3.3	2964	0.53, 0.46

performance. On going from PO-0.01 to PO-0.03 and PO-0.04, the luminous efficiency first rises from 6.2 cd/A to 10.4 cd/A, and then reduces to 8.3 cd/A. The decreased luminous efficiency with Ir complex content higher than 3 mol% may relate to the triplet–triplet (T-T) annihilation [20]. As a result, PO-0.03 has the best device performance among these polymers. A peak luminous efficiency of 10.4 cd/A, a peak power efficiency of 5.0 lm/W, and a maximum brightness of 2842 cd/m^2 have been realized for PO-0.03. Moreover, a gentle roll-off is observed at high current density. For example, at a current density of 20 mA/cm^2, the luminous efficiency can remain as high as 8.0 cd/A. To our knowledge, the performance of PO-0.03 is much better than that of the ever reported yellow PhPs with polyfluorene as the main chain (6.1 cd/A). The result means that the fluorinated poly(arylene ether phosphine oxide)s is a promising platform to construct efficient yellow PhPs.

In conclusion, we report a series of novel partly conjugated yellow-emitting PhPs based on a fluorinated poly(arylene ether phosphine oxide) scaffold. Even at a low Ir content of 2 mol%, the EL emission from FPCzPO is found to be almost quenched completely owing to the efficient intermolecular Förster energy transfer and charge trapping process. The PhPLEDs show optimized luminous efficiency of 10.4 cd/A, higher than that of the reported yellow PhPs. The result suggests that it is promising to adopt this fluorinated poly(arylene ether phosphine oxide) as the platform for the design of high-performance all-phosphorescent single white polymers.

3.4 Experimental Section

Details of measurements, characterization, and fabrication of PLEDs have been described in Chap. 2.

Synthesis FIrpicOH. This compound was synthesized according to the procedure reported in the literature. ^{1}H NMR (300 MHz, $CDCl_3$, δ): 13.53 (br, 1H), 8.67 (d, J = 5.7 Hz, 1H), 8.31 (d, J = 8.5 Hz, 1H), 8.25 (d, J = 8.4 Hz, 1H), 7.81 (t, J = 7.9 Hz, 2H), 7.47 (d, J = 6.6 Hz, 1H), 7.43 (d, J = 5.5 Hz, 1H), 7.29–7.21 (m, 3H), 7.03 (t, J = 6.6 Hz, 1H), 6.54–6.33 (m, 2H), 5.78 (dd, J = 8.7, 2.3 Hz, 1H), 5.57 (dd, J = 8.7, 2.3 Hz, 1H).

1. Bis(4-fluorophenyl)phosphine oxide (10 g, 41.2 mmol), *p*-bromoanisole (9.24 g, 49.0 mmol), *N*-methylmorpholine (6.25 g, 61.8 mmol), and tetrakis (triphenylphosphine)palladium(0) (9.5 g, 8.24 mmol) was added consecutively to 80 mL toluene under argon. The mixture was then heated at 100 °C for 8 h. After cooling to room temperature, the obtained suspension was directly applied to a silica gel column using petroleum ether/ethyl acetate = 1/1 as eluent to give the crude product. Crystallization from a mixture of petroleum ether and CH_2Cl_2 gave the pure product as white crystal in a yield of 88 %. ^{1}H NMR (400 MHz, $CDCl_3$, δ): 7.68–7.61 (m, 4H), 7.55 (dd, J = 11.6, 8.6 Hz, 2H), 7.21–7.11 (m, 4H), 6.98 (dd, J = 8.7, 2.0 Hz, 2H), 3.86 (s, 3H).

2. Into a 50 mL flask were placed **1** (1.38 g, 4 mmol) and dry CH_2Cl_2 (20 mL). The stirred mixture was cooled to −78 °C and then boron tribromide (1.88 ml, 20 mmol) was added. The reaction mixture was warmed to room temperature, stirred for 24 h, and poured into cold water. The mixture was extracted with ethyl acetate and the organic layer was washed with $NaHCO_3$ aqueous and water. After drying, the organic layers were concentrated and applied to a silica gel column using CH_2Cl_2/CH_3OH = 30/1 as eluent to give the product (1.19 g) in a yield of 90 %. ^{1}H NMR (400 MHz, DMSO-d^6, δ): 10.28 (s, 1H), 7.67–7.61 (m, 4H), 7.45–7.33 (m, 6H), 6.92 (dd, J = 8.6, 2.2 Hz, 2H).

3. **2** (2.05 g, 6.21 mmol), 1, 8-dibromooctane (2.88 ml, 18.63 mmol), K_2CO_3 (5 g, 36 mmol), and acetone (30 mL) were added into a 100 mL flask and stirred at flux for 12 h. After cooling to room temperature, the mixture was filtrated and the filter residue was washed with CH_2Cl_2. The combined filtrate was concentrated and chromatographed by silica gel column using petroleum ether/ethyl acetate = 2/3 as eluent to give the product (2.63 g) in a yield of 85 %. ^{1}H NMR (300 MHz, $CDCl_3$, δ): 7.69–7.61 (m, 4H), 7.54 (dd, J = 11.6, 8.7 Hz, 2H), 7.16 (td, J = 8.7, 2.0 Hz, 4H), 6.96 (dd, J = 8.8, 2.2 Hz, 2H), 4.01 (t, J = 6.4 Hz, 2H), 3.43 (t, J = 6.7 Hz, 2H), 1.97–1.73 (m, 4H), 1.60–1.41 (m, 4H).

Blue'-MON. FirpicOH (2.50 g, 3.52 mmol), **3** (2.08 g, 4.22 mmol), K_2CO_3 (10 g, 72 mmol), and DMF (30 mL) were added into a 50 mL flask and stirred at 60 °C for 12 h. After cooling to room temperature, the mixture was extracted with CH_2Cl_2 and the organic layer was washed with water. The organic layers were dried, concentrated, and applied to a silica gel column using ethyl acetate as eluent to give the product (3.44 g) in a yield of 87 %. ^{1}H NMR (300 MHz, $CDCl_3$, δ): 8.80 (d, J = 5.4 Hz, 1H), 8.25 (dd, J = 14.1, 8.9 Hz, 2H), 7.76 (t, J = 7.9 Hz, 2H), 7.69–7.60 (m, 4H), 7.55–7.38 (m, 5H), 7.30–7.25 (m, 1H), 7.21–7.12 (m, 5H), 6.99–6.93 (m, 3H), 6.49–6.38 (m, 2H), 5.80 (d, J = 8.7 Hz, 1H), 5.32 (dd, J = 8.7 Hz, 2.1 Hz, 1H), 4.12–4.03 (m, 2H), 4.01 (t, J = 6.2 Hz, 2H), 2.05–1.91 (m, 2H), 1.91–1.77 (m, 2H), 1.77–1.43 (m, 8H). MALDI-TOF MS: calcd for $C_{54}H_{43}F_6IrN_3O_5P$: 1151.3 found: 1152.3 $[M + H]^+$. Anal. calcd for $C_{54}H_{43}F_6Ir$-N_3O_5P: C, 56.34; H, 3.77; N, 3.65. found: C, 56.14; H, 3.70; N, 3.50.

4. Diethylphosphite (5.85 mL, 45.5 mmol) was added dropwise at 0 °C to a solution of 3, 4, 5-trifluorophenylmagnesium bromide in tetrahydrofuran which was prepared from 3, 4, 5-trifluoro bromobenzene (17.9 mL, 150 mmol), and magnesium (3.96 g, 165 mmol). The mixture was aged for 30 min at 0 °C, then stirred at ambient temperature for 16 h. After that it was cooled again to 0 °C, and 75 mL NH_4Cl aqueous was then added slowly. The mixture was extracted with diethyl ether and the organic phase was washed with $NaHCO_3$ aqueous and brine, then it was dried over Na_2SO_4. After the solvent had been completely removed, the residue was purified by column chromatography on silica gel using petroleum ether/ethyl acetate = 3/1 as eluent to give the product (13.9 g) in 96 % yield. ^{1}H NMR (400 MHz, $CDCl_3$, δ): 8.01 (d, J = 503.6, 1H), 7.39–7.31 (m, 4H).

6F-Cz-MON. **4** (5 g, 16.1 mmol), 9-(4-bromophenyl)-3,6-di-tert-butyl-9*H*-carbazole (5.82 g, 13.4 mmol), *N*-methylmorpholine (2.71 g, 26.8 mmol), and tetrakis(triphenylphosphine)palladium(0) (3.10 g, 2.68 mmol) were added consecutively to 20 mL toluene under argon. The mixture was then heated at 100 °C for 8 h. After cooling to room temperature, the obtained suspension was directly applied to a silica gel column using petroleum ether/ethyl acetate = 6/1 as eluent to give the crude product. Crystallization from a mixture of petroleum ether and CH_2Cl_2 gave the pure product as white crystal (8.0 g) in a yield of 90 %. ^{1}H NMR (400 MHz, $CDCl_3$, δ): 8.13 (d, J = 1.4 Hz, 2H), 7.87–7.77 (m, 4H), 7.48 (dd, J = 8.7, 1.8 Hz, 2H), 7.44 (d, J = 8.7 Hz, 2H), 7.39 (dd, J = 12.6, 6.3 Hz, 4H), 1.46 (s, 18H). ^{13}C NMR (101 MHz, $CDCl_3$, δ): 152.85, 150.29, 144.09, 143.24, 138.19, 133.58, 133.48, 128.34, 127.30, 126.45, 126.32, 124.10, 123.96, 116.52, 109.11, 34.76, 31.90. ^{19}F NMR (376 MHz, $CDCl_3$, δ): −53.43 (d, J = 19.7 Hz, 4F), −75.34 (t, J = 19.6 Hz, 2F). ^{31}P NMR ($CDCl_3$, 298 K, δ): 24.19. MALDI-TOF MS: calcd for $C_{38}H_{32}F_6NOP$: 663.2. found: 664.1 $[M + H]^+$. Anal. calcd for $C_{38}H_{32}F_6NOP$: C, 68.77; H, 4.86; N, 2.11. found: C, 68.53; H, 4.80; N, 2.03.

5. This compound was prepared from **4** and *p*-bromoanisole according to the procedure for the synthesis of **1**. Column chromatography on silica gel using petroleum ether/ethyl acetate = 1/1 as eluent afforded the desired product in 57 % yield. ^{1}H NMR (400 MHz, $CDCl_3$, δ): 7.53 (dd, J = 12 Hz, 8.8 Hz, 2H), 7.31–7.25 (m, 4H), 7.03 (dd, J = 8.8 Hz, 2.4 Hz, 2H), 3.88 (s, 3H).

6. This compound was prepared from **5** according to the procedure for the synthesis of **2**. Column chromatography on silica gel using petroleum ether/ethyl acetate = 1/1 as eluent afforded the desired product in a yield of 87 %. ^{1}H NMR (400 MHz, DMSO-d^6, δ): 10.3 (s, 1H), 7.65–7.58 (m, 4H), 7.47 (dd, J = 12.1, 8.6 Hz, 2H), 6.94 (dd, J = 8.6, 2.4 Hz, 2H).

7. This compound was prepared from **6** and 1, 8-dibromooctane according to the procedure for the synthesis of **3**. Column chromatography on silica gel using petroleum ether/ethyl acetate = 1/1 as eluent afforded the desired product in a yield of 95 %. ^{1}H NMR (400 MHz, $CDCl_3$, δ): 7.51 (dd, J = 12.0, 8.8 Hz, 2H), 7.31–7.25 (m, 4H), 7.01 (dd, J = 8.9, 2.5 Hz, 2H), 4.02 (t, J = 6.5 Hz, 2H), 3.41 (t, J = 6.8 Hz, 2H), 1.92–1.75 (m, 4H), 1.49–1.42 (m, 4H), 1.41–1.30 (m, 4H).

Blue-MON. This compound was prepared from FIrpicOH (0.071 g, 0.1 mmol) and **7** (0.089 g, 0.15 mmol) according to the procedure for the synthesis of Blue'-MON. Column chromatography on silica gel using ethyl acetate as eluent afforded the desired product in a yield of 91 %. ^{1}H NMR (400 MHz, $CDCl_3$, δ): 8.80 (d, J = 5.7 Hz, 1H), 8.24 (dd, J = 19.3, 8.6 Hz, 2H), 7.75 (t, J = 7.9 Hz, 2H), 7.50 (dd, J = 12.0, 8.8 Hz, 2H), 7.45 (dd, J = 13.4, 7.0 Hz, 2H), 7.40 (s, 1H), 7.30–7.24 (m, 5H) 7.18 (t, J = 6.6 Hz, 1H), 7.00 (dd, J = 8.9, 2.5 Hz, 2H), 6.95 (d, J = 6.1 Hz, 1H), 6.50–6.31 (m, 2H), 5.80 (dd, J = 8.7, 2.3 Hz, 1H), 5.52 (dd, J = 8.7, 2.3 Hz,

1H), 4.13–4.05 (m, 2H), 4.00 (t, $J = 6.5$ Hz, 2H), 1.97–1.91 (m, 2H), 1.83–1.76 (m, 2H), 1.63–1.35 (m, 8H). ^{31}P NMR ($CDCl_3$, 298 K, δ): 24.77. MALDI-TOF MS: calcd for $C_{54}H_{39}F_{10}IrN_3O_5P$: 1223.2. found:1224.2 $[M + H]^+$. Anal. calcd for $C_{54}H_{39}F_{10}IrN_3O_5P$: C, 53.03; H, 3.21; N, 3.44. found: C, 52.80; H, 3.18; N, 3.24.

9. Under argon, *n*-BuLi (2.50 M, 2.92 mL) was added dropwise to a solution of 9,9'-diethyl-2-bromofluorene (2.00 g, 6.64 mmol) in 50 mL tetrahydrofuran at −78 °C. After the mixture was aged for 30 min, dry *N,N*-dimethylformamide (1.03 mL, 13.30 mmol) was added slowly, and then the reaction solution was allowed to return to ambient temperature. After stirring for 12 h, 50 mL of water was added to quench the reaction. The mixture was extracted with dichloromethane (30 mL × 3). The combined organic phases were washed with brine, dried over Na_2SO_4, and then concentrated under reduced pressure to give a crude compound. The crude product was purified by column chromatography on silica gel using petroleum ether/dichloromethane = 2/1 as eluent to give the product (1.58 g) in 95 % yield. ^{1}H NMR (300 MHz, $CDCl_3$, δ): 10.06 (s, 1H), 7.89–7.84 (m, 3H), 7.82–7.76 (m, 1H), 7.43–7.34 (m, 3H), 2.18–1.97 (m, 4H), 0.30 (t, $J = 7.4$ Hz, 6H).

10. **9** (1.0 g, 4.0 mmol) and 2-aminodiphenylamine (0.663 g, 3.6 mmol) were dissolved in 40 mL 2-ethoxyethanol. To this solution were added sequentially H_2O_2 (0.86 mL, 28 mmol) and concentrated hydrochloric acid (36 %) (0.44 mL, 14 mmol). After stirring for 8 h, 50 mL of $NaHCO_3$ aqueous was added to neutralize the reaction. The mixture was extracted with dichloromethane and the organic phase was washed with brine and dried over Na_2SO_4. After the solvent had been completely removed, the residue was purified by column chromatography on silica gel using dichloromethane as eluent to give the product (0.67 g) in 45 % yield. ^{1}H NMR (300 MHz, $CDCl_3$, δ): 7.92 (d, $J = 7.7$ Hz, 1H), 7.73 (d, $J = 8.0$ Hz, 1H), 7.70–7.63 (m, 2H), 7.53–7.43 (m, 3H), 7.38–7.25 (m, 9H), 1.96–1.65 (m, 4H), 0.14 (t, $J = 7.3$ Hz, 6H).

11. Under argon, Iridium chloride trihydrate (0.185 g, 0.52 mmol) and **10** (0.50 g, 1.21 mmol) were added to a mixed solvent of 21 mL 2-ethoxyethanol and 7 mL water. The mixture was refluxed for 24 h. After cooling to room temperature, the precipitate was collected by filtration and washed with water and hexane, and the resulting dimer (0.52 g, yield 92 %) was obtained and used directly in the next step.

12. 4-Iodophenol (30 g, 136 mmol), 1, 8-dibromooctane (75 ml, 409 mmol), K_2CO_3 (56 g, 410 mmol), and acetone (300 mL) were added into a 500 mL flask and stirred at flux for 12 h. After cooling to room temperature, the mixture was filtrated and the filter residue was washed with dichloromethane. The combined filtrate was concentrated and distilled under reduced pressure to give the product (53.1 g) as distillate in 95 % yield. ^{1}H NMR (300 MHz, $CDCl_3$, δ): 7.52 (d, $J = 9.0$ Hz, 2H), 6.65 (d, $J = 9.0$ Hz, 2H), 3.89 (t, $J = 6.5$ Hz, 2H), 3.40 (t, $J = 6.7$ Hz, 2H), 1.93–1.64 (m, 4H), 1.50–1.22 (m, 8H).

13. Acetylacetone (3 mL, 29.4 mmol) was added dropwise to a stirred suspension of sodium hydride (3.52 g, 150 mmol) in tetrahydrofuran (90 mL) at 0 °C. The resultant mixture was stirred for 30 min. *n*-BuLi (2.5 M in hexane, 11.2 mL) was added dropwise at 0 °C and the resultant mixture was stirred for a further

30 min at 0 °C. A solution of compound **12** (6.17 g, 15 mmol) in tetrahydrofuran (10 mL) was added dropwise to the stirred solution at 0 °C. The reaction was kept at 0 °C for 1 h and then another 1 h at room temperature, after which it was poured into aqueous ammonium chloride (10 %, 400 mL). The mixture was extracted with dichloromethane. The combined organic phases were washed with brine and dried over anhydrous Na_2SO_4. After removal of the solvent, the crude product was purified by silica gel column chromatography with petroleum ether/ethyl acetate = 1/1 as eluent to give a white solid (5.7 g) in 88 % yield. ^{1}H NMR (300 MHz, $CDCl_3$, δ): 15.49 (s, 0.4H), 7.52 (d, J = 9.0 Hz, 2H), 6.65 (d, J = 9.0 Hz, 2H), 5.47 (s, 0.6H), 3.88 (t, J = 6.5 Hz, 2.3H), 3.55 (s, 0.2H), 3.39 (t, J = 6.8 Hz, 0.6H), 2.48 (t, J = 7.3 Hz, 0.2H), 2.26–2.22 (m, 1.6 H), 2.03 (s, 1.8H), 1.89–1.80 (m,0.6H), 1.80–1.69 (m, 2.4H),1.60–1.50 (m, 2 H), 1.47–1.25 (m, 10H).

14. **4** (4 g, 12.9 mmol), **13** (5.83 g, 13.5 mmol), *N*-methylmorpholine (2.63 g, 25.8 mmol) and tetrakis(triphenylphosphine)palladium(0) (2.98 g, 2.58 mmol) were added consecutively to 20 mL toluene under argon. The mixture was then heated at 95 °C for 3 h. After cooling to room temperature, the obtained suspension was directly applied to a silica gel column using petroleum ether/ethyl acetate = 1/1 as eluent to give the product (6.8 g) in a yield of 86 %. ^{1}H NMR (300 MHz, $CDCl_3$, δ): 15.51 (s, 0.6H), 7.51 (dd, J = 12.0, 8.7 Hz, 2H), 7.32–7.24 (m, 4H), 7.01 (dd, J = 8.7, 2.3 Hz, 2H), 5.49 (s, 0.8H), 4.02 (t, J = 6.5 Hz, 2H), 3.57 (s, 0.3H), 2.50 (t, J = 7.3 Hz, 0.3H), 2.29–2.24(m, 2H), 2.06–2.04 (m, 3.6H), 1.88–1.73 (m, 2H), 1.66–1.52 (m, 2H), 1.52–1.40 (m, 2H), 1.34–1.28 (m, 8H).

Y-MON. The dimer **11** (0.515 g, 0.244 mmol), **14** (0.2991 g, 0.488 mmol), and Na_2CO_3 (0.52 g, 4.88 mmol) were added to a mixed solvent of acetonitrile (5 mL) and chloroform (5 mL) and then refluxed for 24 h. After cooling to room temperature, the obtained suspension was directly applied to a silica gel column (using dichloromethane/ethyl acetate = 100/3 as eluent) and then an alkaline alumina column (using dichloromethane/ethyl acetate = 50/1 as eluent) to give an orange-red solid (0.68 g) in a yield of 85 %. ^{1}H NMR (400 MHz, $CDCl_3$, δ): 7.81 (d, J = 7.4 Hz, 1H), 7.76–7.65 (m, 9H), 7.65–7.56 (m, 1H), 7.51 (dd, J = 12.0, 8.8 Hz, 2H), 7.36–7.20 (m, 12H), 7.15–7.05(m, 7H), 7.00 (dd, J = 8.9, 2.5 Hz, 2H), 6.89 (s, 1H), 6.67 (s, 1H), 6.46 (s, 1H), 6.41 (s, 1H), 5.34 (s, 1H), 3.97 (t, J = 6.5 Hz, 2H), 2.28–2.02 (m, 2H), 1.92 (s, 3H), 1.77–1.70 (m, 2H), 1.63–1.43 (m, 6H), 1.43–1.30 (m, 6H), 1.24–1.15 (m, 2H), 1.09–0.90 (m, 6H), 0.28 (td, J = 7.3, 2.4 Hz, 6H), –0.05 (td, J = 7.2, 2.7 Hz, 6H). ^{31}P NMR ($CDCl_3$, 298 K, δ): 24.75. FTIR (KBr, cm^{-1}): 1580, 1522 (C = O); 1417 (C–P); 1319 (C–F); 1113 (P = O). MALDI-TOF MS: calcd for $C_{91}H_{80}F_6IrN_4O_4P$: 1630.6. found: 1630.6 [M]+. Anal. calcd for $C_{91}H_{80}F_6IrN_4O_4P$: C, 67.02; H, 4.94; N, 3.44. found: C, 66.72; H, 4.90; N, 3.35.

PB'-0.05. Under argon, HO-Cz-MON (0.2938 g, 0.5 mmol), F-Cz-MON (0.2663 g, 0.45 mmol), Blue'-MON (0.0562 g, 0.05 mmol), K_2CO_3 (0.21 g, 1.5 mmol), DMAc (3.0 mL), and toluene (3.0 mL) were added to a 10 mL flask equipped with a magnetic stir bar, a oil-water separator, a condenser, and a gas adapter. The reaction was stirred at 140 °C for 3 h and then 165 °C for 16 h. After that, a solution of F-Cz-MON (0.0295 g, 0.05 mmol) in 1 mL DMAc was added and the mixture was stirred at 165 °C for another 8 h. After cooling to room temperature, the mixture was extracted with CH_2Cl_2. The organic phase was washed with water and then dried over Na_2SO_4. The solution was concentrated and then precipitated in methanol to afford the polymer as white fiber (0.40 g) in a yield of 70 %. ^{1}H NMR (400 MHz, $CDCl_3$, δ): ^{1}H NMR (300 MHz, $CDCl_3$, δ): 8.11 (s, 2H), 7.94–7.64 (m, 8H), 7.48–7.34 (br, 4H), 7.19 (d, $J = 7.2$ Hz, 4H), 4.06–3.91 (m, 0.05H), 1.43 (s, 18H). [Here, similarly hereinafter, only the signals of the $-CH_2O-$ groups of the blue comonomer part are given for clearity].

General Synthetic Procedure for Polymers with the Fluorinated Poly(arylene ether phosphine oxide) Backbone: Under argon, 6F-Cz-MON, HO-Cz-MON, Blue-MON, K_2CO_3 (0.21 g, 1.5 mmol), toluene (2.5 mL), and DMAc (2.5 mL) were added to a 10 mL flask and then stirred at 120 °C for 18 h. After that, a solution of 6F-Cz-MON (10 mol% relative to HO-Cz-MON) in 1 mL DMAc was added and the mixture was stirred at 165 °C for another 8 h. After cooling to room temperature, the mixture was extracted with CH_2Cl_2. The organic phase was washed with water and dried over Na_2SO_4. The solution was concentrated and then precipitated in methanol to afford the polymers in 45 ~ 55 % yield.

FPCzPO: White fiber. 6F-Cz-MON (0.1659 g, 0.25 mmol) and HO-Cz-MON (0.1469 g, 0.25 mmol) were used in the polymerization. ^{1}H NMR (300 MHz, $CDCl_3$, δ): 8.11 (br, 4H), 7.96–7.61 (m, 12H), 7.54–7.34(m, 12H), 7.18–7.95(m, 4H) 1.44 (br, 36H). ^{13}C NMR (100 MHz, $CDCl_3$, δ): 159.77, 156.07 (dd, $J = 259.4$, 18.7 Hz), 144.11, 143.68, 143.27, 142.03, 138.42, 138.18, 134.30, 134.20, 133.60, 130.70, 130.42, 129.64, 129.38, 127.79, 127.20, 126.71, 126.46, 126.33, 126.11, 125.98, 124.12, 123.96, 123.81, 116.52, 116.39, 115.85, 115.72, 109.12, 34.74, 31.91. ^{19}F NMR (376 MHz, $CDCl_3$, δ): −44.75 (s, 1F). ^{31}P NMR ($CDCl_3$, 298 K, δ): 27.04, 24.43. Anal. Calcd: C, 75.36; H, 5.66; N, 2.31. Found: C, 75.00; H, 5.76; N, 2.18.

PB-0.025. Light-green fiber. 6F-Cz-MON (0.2521 g, 0.380 mmol), HO-Cz-MON (0.2351 g, 0.400 mmol), and Blue-MON (0.0245 g, 0.020 mmol) were used in the polymerization. ^{1}H NMR (300 MHz, $CDCl_3$, δ): 8.10 (br, 4H), 7.92–7.61 (m, 12H), 7.52–7.32(m, 12H), 7.16–6.97 (m, 4H), 4.14–3.84 (m, 0.15H), 1.42 (br, 36H). ^{31}P NMR ($CDCl_3$, 298 K, δ): 26.96, 34.35. Anal. Calcd: C, 74.31; H, 5.57; N, 2.37. Found: C, 74.01; H, 5.67; N, 2.21.

PB-0.05. Light-green fiber. 6F-Cz-MON (0.2389 g, 0.360 mmol), HO-Cz-MON (0.2351 g, 0.400 mmol), and Blue-MON (0.0490 g, 0.040 mmol) were used in the polymerization. ^{1}H NMR (300 MHz, $CDCl_3$, δ): 8.11 (br, 4H), 7.95–7.62 (m, 12H), 7.51–7.34(m, 12H), 7.15–6.96 (m, 4H), 4.14–3.84 (m, 0.35H), 1.43 (br, 36H). ^{31}P NMR ($CDCl_3$, 298 K, δ):27.16, 24.37. Anal. Calcd: C, 73.50; H, 5.47; N, 2.43. Found: C, 73.21; H, 5.55; N, 2.38.

PB-0.075. Light-green fiber. 6F-Cz-MON (0.2256 g, 0.340 mmol), HO-Cz-MON (0.2351 g, 0.400 mmol), and Blue-MON (0.0730 g, 0.060 mmol) were used in the polymerization. ^{1}H NMR (300 MHz, $CDCl_3$, δ): 8.11 (br, 4H), 7.93–7.63 (m, 12H), 7.53–7.33(m, 12H), 7.16–6.88 (m, 4H), 4.14–3.84 (m, 0.60H), 1.44 (br, 36H). ^{31}P NMR ($CDCl_3$, 298 K, δ):27.2, 24.40. Anal. Calcd: C, 72.61; H, 5.39; N, 2.48. Found: C, 72.35; H, 5.35; N, 2.36.

PB-0.10. Green fiber. 6F-Cz-MON (0.2123 g, 0.320 mmol), HO-Cz-MON (0.2351 g, 0.400 mmol), and Blue-MON (0.0978 g, 0.080 mmol) were used in the polymerization. ^{1}H NMR (300 MHz, $CDCl_3$, δ): 8.10 (br, 4H), 7.92–7.57 (m, 12H), 7.51–7.26(m, 12H), 7.15–6.85 (m, 4H), 4.14–3.84 (m, 0.79H), 1.43 (br, 36H). ^{31}P NMR ($CDCl_3$, 298 K, δ):27.12, 24.40. Anal. Calcd: C, 71.60; H, 5.30; N, 2.54. Found: C, 71.32; H, 5.36; N, 2.35.

PY-0.01. Yellow fiber (0.28 g). Yield: 56 %. 6F-Cz-MON (0.2601 g, 0.392 mmol), HO-Cz-MON (0.2351 g, 0.400 mmol), and Y-MON (0.0130 g, 0.008 mmol) were used in the polymerization. ^{1}H NMR (300 MHz, $CDCl_3$, δ): 8.11 (s, 4H), 7.96–7.61 (m, 12H), 7.54–7.34 (m, 12H), 7.18–7.95 (m, 4H) 1.44 (s, 36H). FTIR (KBr, cm^{-1}): 1413 (C–P); 1319 (C–F); 1234 (C–O–C); 1115 (P = O). Anal. calcd: C, 75.22; H, 5.66; N, 2.34. found: C, 75.02; H, 5.80; N, 2.29.

PO-0.02. Yellow fiber (0.26 g). Yield: 51 %. 6F-Cz-MON (0.2548 g, 0.384 mmol), HO-Cz-MON (0.2351 g, 0.400 mmol), and Y-MON (0.0261 g, 0.016 mmol) were used in the polymerization. ^{1}H NMR (300 MHz, $CDCl_3$, δ): 8.10 (s, 4H), 7.96–7.61 (m, 12H), 7.54–7.34(m, 12H), 7.18–7.95(m, 4H) 1.44 (s, 36H), 0.30–0.20 (m, 0.20H). FTIR (KBr, cm^{-1}): 1414 (C–P); 1319 (C–F); 1236 (C–O–C); 1117 (P = O). Anal. calcd: C, 75.10; H, 5.65; N, 2.38. found: C, 74.88; H, 5.79; N, 2.25.

PO-0.03. Yellow fiber (0.27 g). Yield: 55 %. 6F-Cz-MON (0.2495 g, 0.376 mmol), HO-Cz-MON (0.2351 g, 0.400 mmol), and Y-MON (0.0391 g, 0.024 mmol) were used in the polymerization. ^{1}H NMR (300 MHz, $CDCl_3$, δ): 8.10 (s, 4H), 7.96–7.61 (m, 12H), 7.54–7.34 (m, 12H), 7.18–7.95 (m, 4H) 1.44 (s, 36H), 0.28–0.20 (m, 0.43H). FTIR (KBr, cm^{-1}): 1413 (C–P); 1319 (C–F); 1236 (C–O–C); 1117 (P = O). Anal. calcd: C, 74.98; H, 5.65; N, 2.41. found: C, 74.76; H, 5.90; N, 2.38 .

PO-0.04. Yellow fiber (0.26 g). Yield: 50 %. 6F-Cz-MON (0.2442 g, 0.368 mmol), HO-Cz-MON (0.2351 g, 0.400 mmol), and Y-MON (0.0522 g, 0.032 mmol) were used in the polymerization. ^{1}H NMR (300 MHz, $CDCl_3$, δ): 8.11 (s, 4H), 7.96–7.61 (m, 12H), 7.54–7.34 (m, 12H), 7.18–7.95 (m, 4H) 1.44 (s, 36H). FTIR (KBr, cm^{-1}): 1414 (C–P); 1319 (C–F); 1236 (C–O–C); 1117 (P = O). Anal. calcd: C, 74.87; H, 5.64; N, 2.44. found: C, 74.50; H, 5.67; N, 2.33

References

1. Chen XW, Liao JL, Liang YM et al (2003) High-efficiency red-light emission from polyfluorenes grafted with cyclometalated iridium complexes and charge transport moiety. J Am Chem Soc 125:636–637
2. Jiang JX, Jiang CY, Yang W et al (2005) High-efficiency electrophosphorescent fluorene-alt-carbazole copolymers n-grafted with cyclometalated ir complexes. Macromolecules 38:4072–4080
3. Evans NR, Devi LS, Mak CSK et al (2006) Triplet energy back transfer in conjugated polymers with pendant phosphorescent iridium complexes. J Am Chem Soc 128:6647–6656
4. Zhen H, Luo J, Yang W et al (2007) Novel light-emitting electrophosphorescent copolymers based on carbazole with an ir complex on the backbone. J Mater Chem 17:2824–2831
5. Chien CH, Liao SF, Wu CH et al (2008) Electrophosphorescent polyfluorenes containing osmium complexes in the conjugated backbone. Adv Funct Mater 18:1430–1439

6. Yang X-H, Wu F-I, Neher D et al (2008) Efficient red-emitting electrophosphorescent polymers. Chem Mater 20:1629–1635
7. Ma ZH, Ding JQ, Zhang BH et al (2010) Red-emitting polyfluorenes grafted with quinoline-based iridium complex: "simple polymeric chain, unexpected high efficiency". Adv Funct Mater 20:138–146
8. Ma Z, Chen L, Ding J et al (2011) Green electrophosphorescent polymers with poly(3,6-carbazole) as the backbone: a linear structure does realize high efficiency. Adv Mater 23:3726–3729
9. Tokito S, Suzuki M, Sato F et al (2003) High-efficiency phosphorescent polymer light-emitting devices. Org Electron 4:105–111
10. Fei T, Cheng G, Hu D et al (2010) Iridium complex grafted to 3,6-carbazole-alt-tetraphenylsilane copolymers for blue electrophosphorescence. J Polym Sci, Part A: Polym Chem 48:1859–1865
11. Zhen HY, Jiang CY, Yang W et al (2005) Synthesis and properties of electrophosphorescent chelating polymers with iridium complexes in the conjugated backbone. Chemistry-a Europ J 11:5007–5016
12. Wu HB, Zhou GJ, Zou JH et al (2009) Efficient polymer white-light-emitting devices for solid-state lighting. Adv Mater 21:4181–4184
13. Zou J, Wu H, Lam C-S et al (2011) Simultaneous optimization of charge-carrier balance and luminous efficacy in highly efficient white polymer light-emitting devices. Adv Mater 23:2976–2980
14. Sudhakar M, Djurovich PI, Hogen-Esch TE et al (2003) Phosphorescence quenching by conjugated polymers. J Am Chem Soc 125:7796–7797
15. Shao S, Ding J, Wang L et al (2012) Highly efficient blue electrophosphorescent polymers with fluorinated poly(arylene ether phosphine oxide) as backbone. J Am Chem Soc 134:15189–15192
16. Shao S, Ding J, Wang L et al (2012) Synthesis and characterization of yellow-emitting electrophosphorescent polymers based on a fluorinated poly(arylene ether phosphine oxide) scaffold. J Mater Chem 22:24848–24855
17. Bernal DP, Bankey N, Cockayne RC et al (2002) Fluoride-terminated hyperbranched poly (arylene ether phosphine oxide)s via nucleophilic aromatic substitution. J Polym Sci, Part A: Polym Chem 40:1456–1467
18. Nielsen KT, Bechgaard K, Krebs FC (2005) Removal of palladium nanoparticles from polymer materials. Macromolecules 38:658–659
19. Forster T (1959) 10th spiers memorial lecture—transfer mechanisms of electronic excitation. Discuss Faraday Soc 27:7–17
20. Gong X, Ma WL, Ostrowski JC et al (2004) White electrophosphorescence from semiconducting polymer blends. Adv Mater 16:615–619

Chapter 4
All-Phosphorescent Single-Component White Polymers

4.1 Background

White polymer light-emitting diodes (WPLEDs) [1] have received great attention because of their potential application in backlight for flat panel displays and solid-state lighting sources. Basically, there are two approaches to realize high-efficiency WPLEDs. One is the polymer blend system [2–4] based on the additive color-mixing theory where the blend of the three primary colors (red, green, and blue) or two complementary colors (blue and orange) lead to pure white emission. However, voltage dependence of emitted color and phase separation hinder the utilization of the polymer blends [5]. To overcome these issues, an alternative approach, i.e., white-light emission from a single-component system, is desirable. Several studies have succeeded in realizing highly efficient WPLEDs based on polyfluorene derivatives containing red-, green-, and blue-emitting units in the backbone or the sidechain [6]. The color of the emission can be finely tuned by adjusting the composition of the RGB units and controlling the partial energy transfer from blue- to the low-energy-emissive units.

According to the nature of the dopants, single-component white polymers (SWPs) can be divided into three classes (taking two-color SWPs as example, see Fig. 4.1): (I) all-fluorescent SWPs with singlet blue and yellow emitters [6, 12, 13]; (II) fluorescent/phosphorescent hybrid SWPs with singlet/triplet blue and yellow emitters [9, 11]; (III) all-phosphorescent SWPs with triplet blue and yellow emitters [14]. Early in 2004, our group reported the all-fluorescent SWPs with a luminous efficiency of 5.3 cd/A with CIE coordinates of (0.25, 0.35) [6]. In order to utilize both singlet and triplet excitons in the electroluminescent process, fluorescent/phosphorescent hybrid SWPs were developed by Cao's group with a luminous efficiency of 4.5 cd/A and CIE coordinates of (0.44, 0.32) in 2007 [9]. Since then numerous works have been devoted to the design of all-fluorescent and fluorescent/phosphorescent hybrid SWPs, and their reported highest efficiencies have reached to 12.8 [13] and 10.7 cd/A [11], respectively, corresponding to external quantum

S. Shao, *Electrophosphorescent Polymers Based on Polyarylether Hosts*,
Springer Theses, DOI 10.1007/978-3-662-44376-7_4

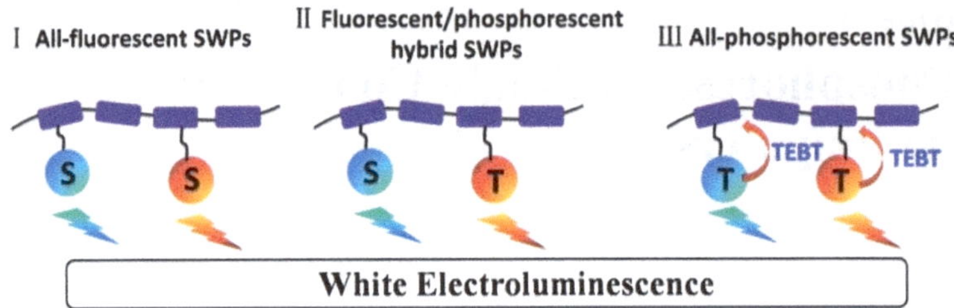

Fig. 4.1 Schematic diagram of three kinds of two-color SWPs. S and T represent singlet and triplet emitters, respectively. Reprinted with permission from Ref. [15]. Copyright 2012 American Chemical Society

efficiencies (EQEs) of about 5 %. Obviously, the relatively low EQEs indicate that the exictons are still not utilized completely. In this context, designing all-phosphorescent SWPs where all the emitters can harvest both singlet and triplet excitons are highly desirable.

However, the development of all-phosphorescent SWPs remains challenging, and only one example showing an EQE of only 1.5 % has been reported till date [14]. As the performance of SWPs is dependent on both the blue and yellow emitting segments, the problems existing in blue/yellow PhPs also exist in SWPs. In particular, the scarcity of suitable polymer hosts with high triplet energy above 2.70 eV as well as matched HOMO/LUMO levels with the Fermi levels of the electrodes has been recognized as the main reason for the development bottleneck of SWPs.

4.2 Molecular Designs

In Chap. 2, we have demonstrated that a trade-off between triplet energy and HOMO/LUMO levels are realized for the polyarylether hosts. Consequently, in Chap. 3, the fluorinated poly (arylene ether phosphine oxide) (FPCzPO) scaffold is successfully used in the construction of blue and yellow PhPs. Therefore, for the design of all-phosphorescent SWPs, we believe that FPCzPO will also be a promising platform. In this chapter, we graft blue and yellow phosphors simultaneously to the sidechain of FPCzPO to construct all-phosphorescent SWPs [15]. The molecular structures of the SWPs are shown in Fig. 4.2. Here FIrpic and $(fbi)_2Ir$ (acac) are selected as the blue and yellow phosphors, respectively. The content of the blue and yellow dopants are adjusted to control the energy transfer process to achieve white emission.

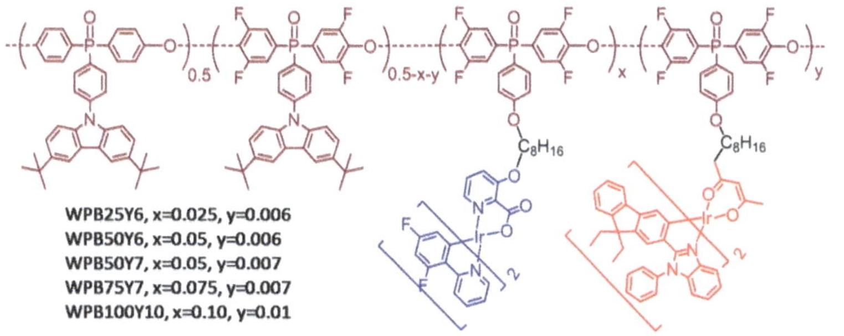

Fig. 4.2 Molecular structures of all-phosphorescent SWPs. Reprinted with permission from Ref. [15]. Copyright 2012 American Chemical Society

4.3 Results and Discussions

Synthesis and characterization similar low temperature polycondensation (120 °C) condition used for PB-0.05 is adopted for all-phosphorescent SWPs. However, the synthesis of these SWPs is still full of challenge, for it seems problematic when both Blue-MON and Y-MON are added together with HO-Cz-MON and 6F-Cz-MON at the beginning of polymerization ("one-step addition" procedure shown in Scheme 4.1). As indicated in Fig. 4.3a, the EL emission from the yellow part in WP'-B100-Y10 displays a hypsochromic shift of ~10 nm compared with the yellow PhP PO-0.01 described in Chap. 3. This abnormity may be caused by the ancillary ligand exchange between Blue-MON and Y-MON, as the replacement of acetylacetonate ligand with picolinate would lead to a similar blue-shift [16, 17]. Further experiments should be performed to verify this hypothesis, but they are now beyond the aim of this study.

To avoid undesired color change, **Blue-MON** is first copolymerized with **HO-Cz-MON** and **6F-Cz-MON**, and successively the other complex monomer **Y-MON** is added to continue the polymerization (See the "two-step addition" strategy as shown in Scheme 4.2). With this modified method, a series of all-phosphorescent SWPs (WPB25Y6 ~ WPB100Y10) with different Ir loadings are successfully prepared. It should be noted that, unlike WP'-B100-Y10, all the polymers show normal yellow emission from $(fbi)_2Ir(acac)$, which matches well with that of PO-0.01 (Fig. 4.3b).

The number-average molecular weight (M_n) and the polydispersity index (PDI) of the polymers were measured by gel permeation chromatography (GPC), and the data are listed in Table 4.1. Moderate molecular weights of 68,00 ~ 7,700 Da for all the SWPs are obtained, and they do not alter obviously with the Ir loading. The polymers exhibit good solubility in common organic solvents, such as dichloromethane, tetrahydrofuran, toluene, and chlorobenzene. Thermogravimetric analysis (TGA) indicates that they are thermally stable with the decomposition temperatures (T_d) higher than 370 °C.

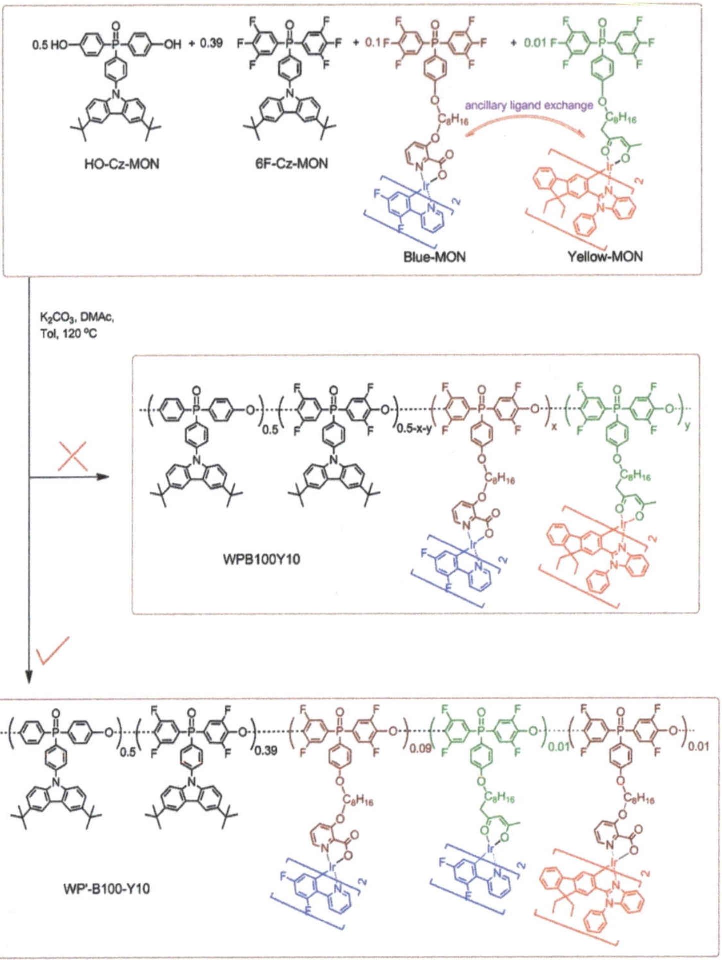

Scheme 4.1 "One-step addition" routes for the synthesis of all-phosphorescent SWPs

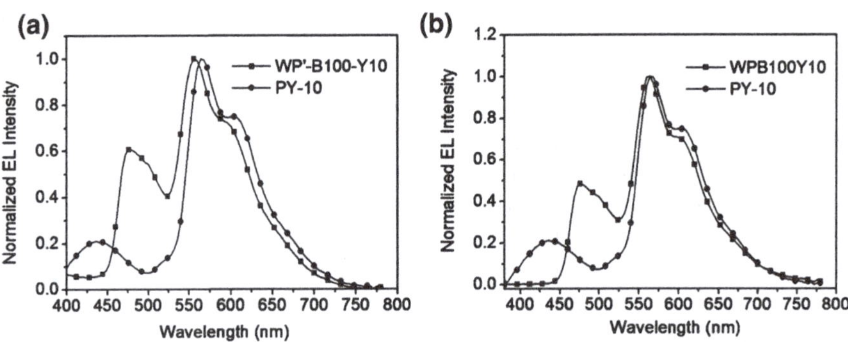

Fig. 4.3 Comparison of the EL spectra of WP'-B100-Y10 (**a**) and WPB100Y10 (**b**) with that of PY-10

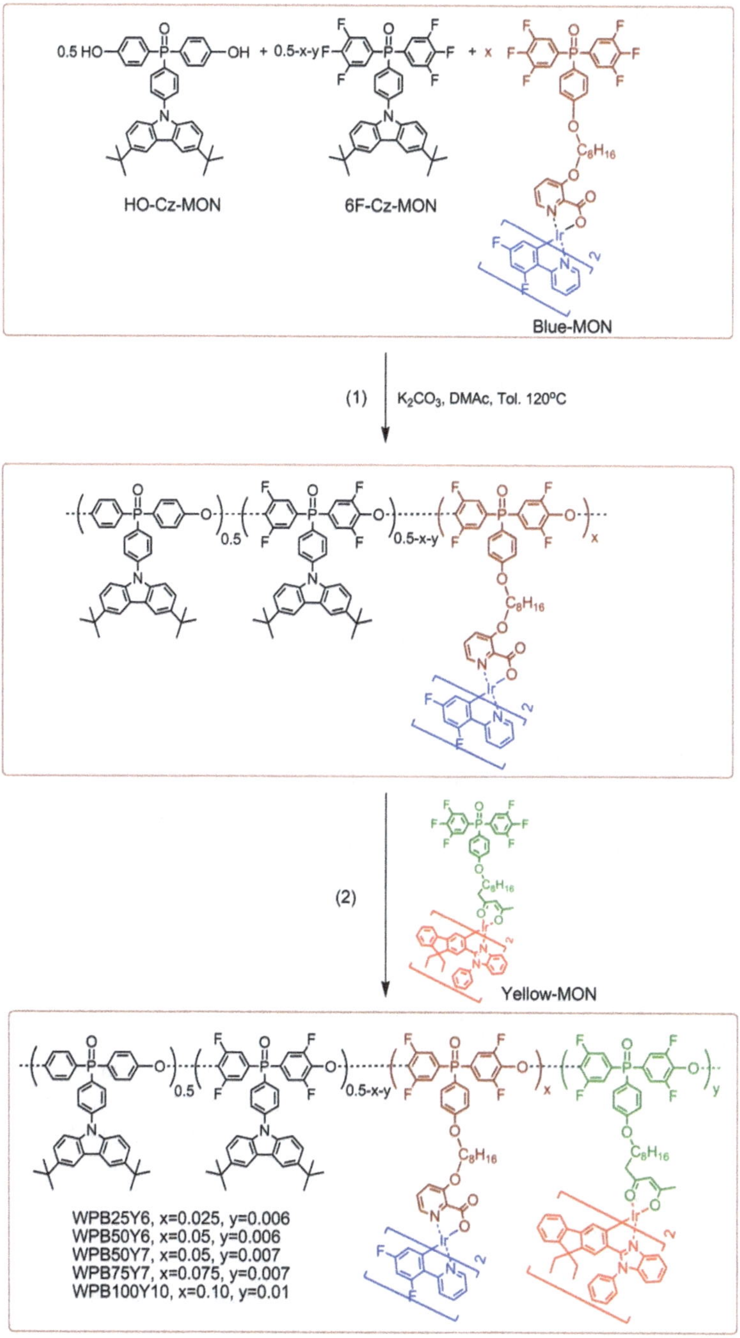

0.5 HO-Cz-MON + 0.5-x-y 6F-Cz-MON + x Blue-MON

(1) K_2CO_3, DMAc, Tol. 120°C

(2) Yellow-MON

WPB25Y6, x=0.025, y=0.006
WPB50Y6, x=0.05, y=0.006
WPB50Y7, x=0.05, y=0.007
WPB75Y7, x=0.075, y=0.007
WPB100Y10, x=0.10, y=0.01

Scheme 4.2 "Two-step addition" routes for the synthesis of all-phosphorescent SWPs

Table 4.1 Characterization data of WPB25Y6 ~ WPB100Y10

Polymer	M_n	PDI	T_d	Ir-complex loading (mol%)	
				Feed ratio	Calculated
WPB25Y6	7,500	1.71	419	0.025	0.018
WPB50Y6	7,500	1.74	403	0.050	0.048
WPB50Y7	7,000	1.72	389	0.050	0.041
WPB75Y7	7,700	1.67	380	0.075	0.068
WPB100Y10	6,800	1.76	374	0.10	0.094

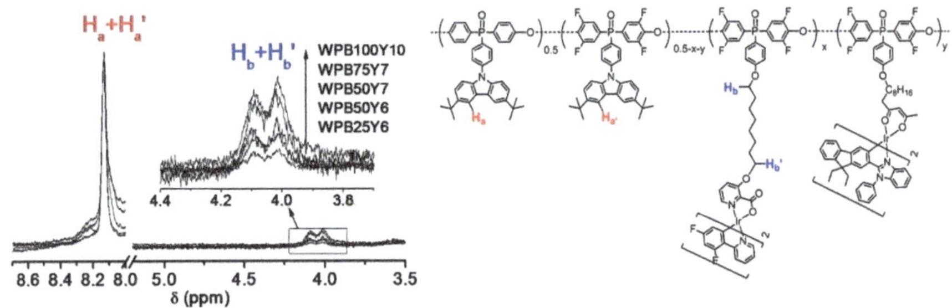

Fig. 4.4 Characteristic peaks of the ^{1}H NMR spectra of the SWPs

The Ir loadings of the polymers are controlled according to the feed ratio, and the actual Ir content in the polymers is calculated through the ^{1}HNMR spectra (Fig. 4.4). The singlet peak at 8.10 ppm is assigned to the 4- and 5-position protons of carbazole units, while the signals at 3.95–4.15 ppm come from –CH_2O– groups of the alkyl spacer in **Blue-MON** (The signals from **Y-MON** can be omitted due to its rather low loading). In this context, the actual content of FIrpic incorporated into the polymer can be quantified by comparing the integration of the singlet peak at 8.10 ppm with those at 3.95–4.15 ppm, and the data are listed in Table 4.1. It is found that the actual FIrpic content is slightly lower than the feed ratio.

Optical properties The absorption and photoluminescent (PL) spectra in solutions for the polymers WPB25Y6 ~ WPB100Y10 are given in Fig. 4.5. All the SWPs exhibit nearly the same absorption spectra with peaks at 296/333/343 nm in dichloromethane. The metal-to-ligand charge transfer (MLCT) transition from the Ir complexes is too weak to be discernible due to their low loadings. The PL spectra of the polymers in toluene show two strong peaks at about 410 nm and 472 nm accompanied with a shoulder at 568 nm, which can be ascribed to emissions from FPCzPO, FIrpic, and $(fbi)_2Ir(acac)$, respectively. In addition, the emission intensity of FIrpic and $(fbi)_2Ir(acac)$ increases gradually when the contents of the Ir complexes increase. This observation indicates that efficient intramolecular Förster energy transfer occurs from the polymer backbone to the Ir complexes, as supported by the good overlap between the PL spectrum of FPCzPO with the absorption spectra of FIrpic and $(fbi)_2Ir(acac)$ as shown in Chap. 3.

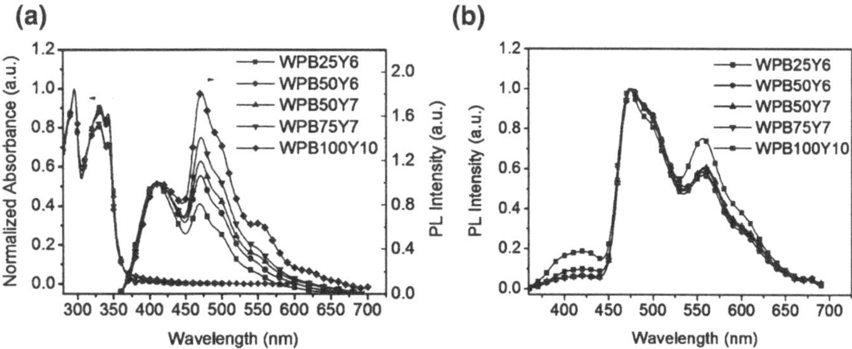

Fig. 4.5 Absorption (in dichloromethane) and PL spectra (in toluene) of the SWPs (**a**), as well as their PL spectra in films (**b**). Reprinted with permission from Ref. [15]. Copyright 2012 American Chemical Society

The polymer films show very different PL profiles from those in solution, the emissions from both FIrpic and $(fbi)_2Ir(acac)$ become dominant in the film state, and the FPCzPO emission almost disappears (Fig. 4.5b). This observation implies that both the intra- and intermolecular energy transfer exist in films. Moreover, we note that, for all the SWPs, the relative intensity ratio of the yellow to blue emission is much higher in film than that in solution. According to the results in Chap. 3, the energy transfer from FPCzPO to $(fbi)_2Ir(acac)$ is not as effective as FIrpic. Therefore, besides the direct energy transfer from FPCzPO to $(fbi)_2Ir(acac)$, we speculate that a cascade process from FPCzPO to FIrpic and then from FIrpic to $(fbi)_2Ir(acac)$ also contributes to the enhancement of the yellow emission. This process is probable since the distance between FIrpic and $(fbi)_2Ir(acac)$ is significantly reduced on going from solution to solid states and therefore the Dexter energy transfer from FIrpic to $(fbi)_2Ir(acac)$ is considerably favored.

Electroluminescent properties To investigate the EL performance of the SWPs, white PhPLEDs have been fabricated with a double-layer device configuration of ITO/PEDOT:PSS (40 nm)/SWPs (40 nm)/TPCz (50 nm)/LiF (1 nm)/Al (100 nm) (Fig. 4.6). Figure 4.7 depicts the EL spectra of WPB75Y7 at different applied

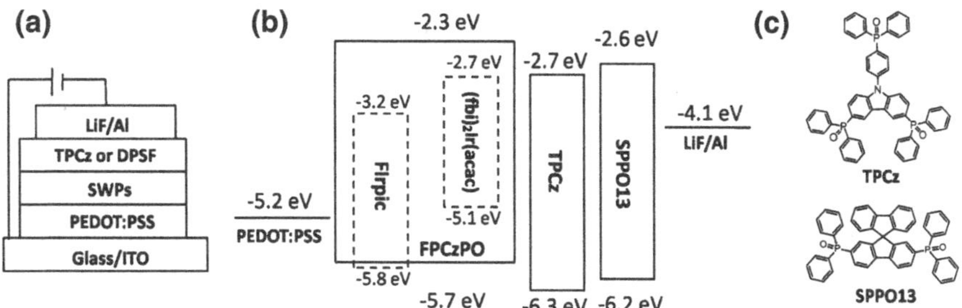

Fig. 4.6 Device configurations (**a**) and the relative energy levels of the materials used for the devices (**b**), as well as the chemical structures of TPCz and SPPO13 (**c**)

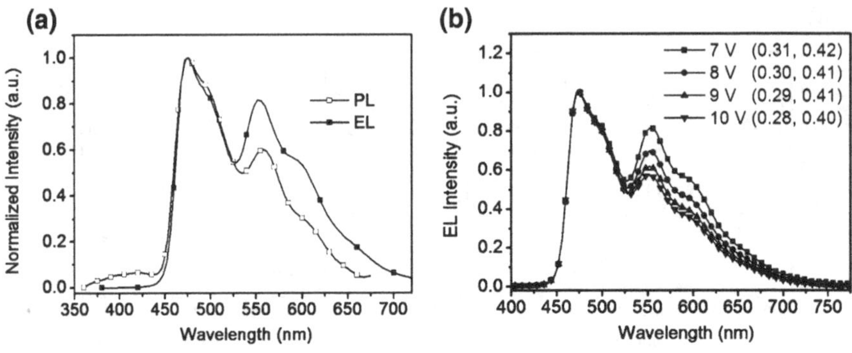

Fig. 4.7 Comparison between the PL and EL spectra (**a**) and EL spectra at different voltages (**b**) for WPB75Y7. Reprinted with permission from Ref. [15]. Copyright 2012 American Chemical Society

voltages and the comparison between its PL and EL spectra. We can see that WPB75Y7 shows white EL with two emission peaks at 472 nm and 568 nm, which come from FIrpic and $(fbi)_2Ir(acac)$, respectively. The EL is mildly bias-independent, with the CIE coordinates slightly shifting from (0.31, 0.42) to (0.28, 0.40) as the driving voltage changes from 7 to 10 V. We note that the bias-independence of the EL spectra is caused by the decline in the yellow emission from $(fbi)_2Ir(acac)$ with the increase in voltage, which can be further attributed to the strong charge trapping effect of $(fbi)_2Ir(acac)$ at low driving voltage. This charge trapping mechanism is consistent with the HOMO/LUMO levels of $(fbi)_2Ir(acac)$ and FPCzPO (Fig. 4.6b), and further proved by the difference between the PL and EL spectra of WPB75Y7. As presented in Fig. 4.7a, the intensity of the $(fbi)_2Ir(acac)$ emission relative to FIrpic in the EL spectrum becomes stronger than that in the PL spectrum, which is indicative of the existence of charge trapping process.

Together with the above-mentioned exciton formation pathway, three possible processes may be involved in the EL of these SWPs: (1) direct Förster energy transfer from FPCzPO to FIrpic and $(fbi)_2Ir(acac)$; (2) direct charge trapping on $(fbi)_2Ir(acac)$; (3) Dexter energy transfer from FIrpic to $(fbi)_2Ir(acac)$. With the synergistic effect of these processes, simultaneous emissions from FIrpic to $(fbi)_2Ir(acac)$ are realized to create white EL.

Figure 4.8 shows the *J-V-B* curves of the SWPs, and their device data are summarized in Table 4.2. First, we note that the efficiency and CIE coordinates are not sensitive to the contents of Ir complexes. For instance, when the feed ratios of the blue and yellow dopant are tuned from 2.5 to 10 mol% and from 0.6 to 1 mol%, respectively, the luminous efficiencies slightly vary within the range of 14.0 ± 1.2 cd/A with CIE coordinates in the range of (0.33 ± 0.05, 0.41 ± 0.02). Secondly, the typical doping concentration in these SWPs (0.6 ~ 10 mol%) is about one to two orders of magnitude higher than that in previous reported all-fluorescent SWPs (0.01 ~ 0.1 mol%). This feature is beneficial for decreasing the batch-to-batch variation in the synthesis of SWPs and thus enhancing the reliability and reproducibility during the fabrication of WPLEDs.

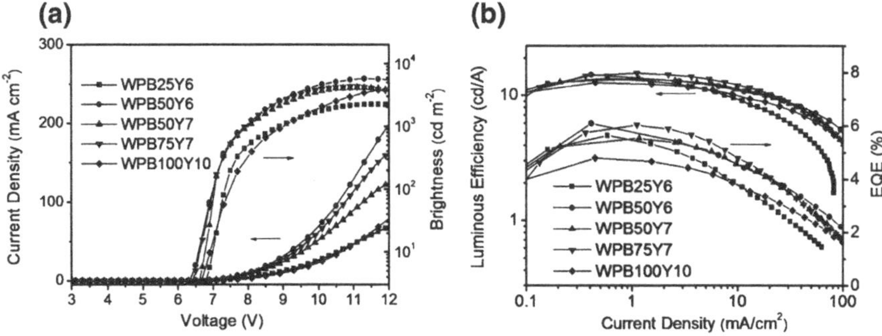

Fig. 4.8 The *J-V–B* (**a**) and η_l-*J*-EQE (**b**) curves of the devices using TPCz as the electron transporting material. Reprinted with permission from Ref. [15]. Copyright 2012 American Chemical Society

Table 4.2 Device performance of single white polymers. Reprinted with permission from Ref. [15]

Polymer	V_{on} (V)	$\eta_{l,\ max}$ (cd A^{-1})	$\eta_{p,\ max}$ (lm W^{-1})	EQE (%)	L_{max} (cd m^{-2})	CIE (x, y)
WPB25Y6[a]	6.5	13.8	6.0	5.6	2,243	(0.32, 0.39)
WPB50Y6[a]	6.1	14.8	6.7	6.1	5,782	(0.28, 0.39)
WPB50Y7[a]	6.3	13.7	6.0	5.5	4,194	(0.36, 0.41)
WPB75Y7[a]	6.1	15.2	6.7	6.0	4,331	(0.30, 0.41)
WPB100Y10[a]	6.3	12.7	5.5	4.8	4,406	(0.37, 0.43)
WPB75Y7[b]	5.8	18.4	8.5	7.1	5,100	(0.31, 0.43)

[a] TPCz is used as the ETM

[b] SPPO13 is used as the ETM

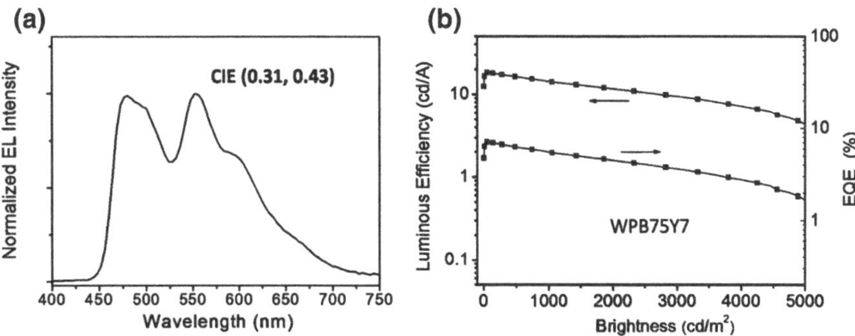

Fig. 4.9 EL spectra (**a**) and η_l-*J*-EQE (**b**) curves of the devices using SPPO13 as the electron transporting material. Reprinted with permission from Ref. [15]. Copyright 2012 American Chemical Society

Among WPB25Y6 ~ WPB100Y10, WPB75Y7 possesses the best device performance with a maximum η_l of 15.2 cd/A, a maximum η_p of 6.7 lm/W, and a peak EQE of 6.0 %. Here the η_p is somewhat low because of the high driving voltage, which is understandable considering that the hole injection barrier from the anode to the SWPs is still as high as ~0.5 eV. This problem may be overcome by further elevating the HOMO levels of the host through molecular structure optimization or inserting a hole injection layer between the anode and the emissive layer in our future work.

When the ETM is changed from TPCz to 9,9′-spirobi(fluorine)-2,7-diylbis (diphenyl phosphine oxide) (SPPO13), a further improved efficiency of 18.4 cd/A (8.5 lm/W, 7.1 %) has been obtained (see Table 4.2 and Fig. 4.9). Even at a brightness of 1,000 cd/m^2, the η_l still remains as high as 14.2 cd/A, indicating a slow efficiency roll-off at high current density. To the best of our knowledge, this is the highest efficiency ever reported for all-phosphorescent SWPs. By comparison to the phosphorescent bichromophoric block copolymers with non-conjugated polystyrene backbone (EQE = 1.5 %) [14], the efficiency is improved by about 4.7 fold. Furthermore, as shown in Table 4.3, it is also much better than those of all-fluorescent and fluorescent/phosphorescent hybrid SWPs, shedding light on the significance of all-phosphorescent SWPs.

In conclusion, based on the fluorinated poly(arylene ether phosphine oxide) backbone that has high triplet energy and appropriate HOMO/LUMO levels, we have synthesized a series of all-phosphorescent SWPs via a "two-step addition" strategy. By tuning the contents of FIrpic and $(fbi)_2Ir(acac)$ to control the energy transfer process, individual blue and yellow emissions are generated to result in standard white electroluminescence with a prominent efficiency as high as 18.4 cd/A. We believe that these all-phosphorescent SWPs bear a promising potential for further performance improvement and the results of this work have opened a window for the design of efficient white-light-emitting polymers in the future.

Table 4.3 Best device performance of different classes of SWPs

Classes of SWPs	CIE (x, y)	LE (cd A^{-1})	EQE (%)	PE (lm W^{-1})	Reference
All-fluorescent SWPs	0.31, 0.36	12.8	5.4	8.5	Liu et al. [13]
	0.33, 0.36	8.6	–	5.4	Liu et al. [10]
Fluorescent/ phosphorescent hybrid SWPs	0.37, 0.30	10.7	5.4	–	Chien et al. [11]
	0.35, 0.38	8.2	3.7	7.2	Wu et al. [18]
	0.32, 0.44	6.1	–	–	Jiang et al. [9]
All-phosphorescent SWPs	0.31, 0.43	18.4	7.1	8.5	This work

4.4 Experimental Section

Detailed information about measurements, characterization, and fabrication of polymer light-emitting diodes can be found in Chap. 3.

Synthesis "One-step addition" polymerization procedure: Under argon, HO-Cz-MON (0.1763 g, 0.300 mmol), 6F-Cz-MON (0.1553 g, 0.234 mmol), Blue-MON (0.0734 g, 0.060 mmol), Yellow-MON (0.0098 g, 0.006 mmol), K_2CO_3 (0.21 g, 1.5 mmol) together with *N*, *N*-dimethylacetamide (DMAc, 2.5 mL) and toluene (2.5 mL) were added to a 10 mL flask and then stirred at 120 °C for 24 h. After cooling to room temperature, the mixture was extracted with CH_2Cl_2. The organic phase was washed with water and dried over Na_2SO_4. After concentration, the organic phase was precipitated in methanol twice to afford the polymers in 48 % yield.

"Two-step addition" polymerization procedure: Under argon, HO-Cz-MON, 6F-Cz-MON, Blue-MON, K_2CO_3 (0.21 g, 1.5 mmol) together with *N*, *N*-dimethyl-acetamide (DMAc, 2 mL), and toluene (2.5 mL) were first added to a 10 mL flask and then stirred at 120 °C for 21 h. Then Yellow-MON dissolved in 0.5 mL *N*, *N*-dimethylacetamide was injected into the flask under argon to continue the polymerization. After stirring for another 3 h, the mixture was cooled to room temperature and extracted with CH_2Cl_2. The organic phase was washed with water and dried over Na_2SO_4. After concentration, the organic phase was precipitated in methanol twice to afford the polymers in 45 ~ 55 % yields.

WPB25Y6. Light-yellow fiber. Yield 55 %. HO-Cz-MON (0.1763 g, 0.300 mmol), 6F-Cz-MON (0.1867 g, 0.2814 mmol), Blue-MON (0.0183 g, 0.015 mmol) and Yellow-MON (0.0059 g, 0.0036 mmol) were used in the polymerization. ^{1}H NMR (300 MHz, $CDCl_3$, δ): 8.10 (br, 4H), 7.92–7.59 (m, 12H), 7.51–7.32 (m, 12H), 7.15–6.97 (m, 4H), 4.19–3.89 (m, 0.14H), 1.43 (br, 36H). ^{31}P NMR ($CDCl_3$, 298 K, δ): 27.00, 24.40.

WPB50Y6. Yellow fiber. Yield 45 %. HO-Cz-MON (0.1763 g, 0.3000 mmol), 6F-Cz-MON (0.1768 g, 0.2664 mmol), Blue-MON (0.0367 g, 0.0300 mmol), and Yellow-MON (0.0059 g, 0.0036 mmol) were used in the polymerization. ^{1}H NMR (300 MHz, $CDCl_3$, δ): 8.14 (br, 4H), 7.97–7.62 (m, 12H), 7.59–7.34 (m, 12H), 7.21–6.83 (m, 4H), 4.17–3.93 (m, 0.39H), 1.47 (br, 36H). ^{31}P NMR ($CDCl_3$, 298 K, δ): 27.16, 24.38.

WPB50Y7. Yellow fiber. Yield 50 %. HO-Cz-MON (0.1763 g, 0.3000 mmol), 6F-Cz-MON (0.1764 g, 0.2658 mmol), Blue-MON (0.0367 g, 0.0300 mmol), and Yellow-MON (0.0068 g, 0.0042 mmol) were used in the polymerization. ^{1}H NMR (300 MHz, $CDCl_3$, δ): 8.10 (br, 4H), 7.94–7.62 (m, 12H), 7.51–7.31 (m, 12H), 7.16–6.93 (m, 4H), 4.18–3.85 (m, 0.33H), 1.47 (br, 36H). ^{31}P NMR ($CDCl_3$, 298 K, δ): 27.01, 24.37.

WPB75Y7. Yellow fiber. Yield 45 %. HO-Cz-MON (0.2351 g, 0.4000 mmol) 6F-Cz-MON (0.2219 g, 0.3344 mmol), Blue-MON (0.0734 g, 0.0600 mmol), and Yellow-MON (0.0091 g, 0.0056 mmol) were used in the polymerization. ^{1}H NMR (300 MHz, $CDCl_3$, δ): 8.10 (br, 4H), 7.93–7.59 (m, 12H), 7.49–7.31 (m, 12H),

7.13–6.90 (m, 4H), 4.15–3.81 (m, 0.54H), 1.43 (br, 36H). ^{31}P NMR ($CDCl_3$, 298 K, δ): 27.20, 24.40.

WPB100Y10. Yellow fiber. Yield 43 %. HO-Cz-MON (0.1763 g, 0.300 mmol), 6F-Cz-MON (0.1553 g, 0.234 mmol), Blue-MON (0.0734 g, 0.060 mmol), Yellow-MON (0.0098 g, 0.006 mmol) were used in the polymerization. ^{1}H NMR (300 MHz, $CDCl_3$, δ): 8.13 (br, 4H), 7.95–7.64 (m, 12H), 7.54–7.36 (m, 12H), 7.17–6.99 (m, 4H), 4.21–3.82 (m, 0.77H), 1.44 (br, 36H). ^{31}P NMR ($CDCl_3$, 298 K, δ): 27.00, 24.38.

References

1. Wu H, Ying L, Yang W et al (2009) Progress and perspective of polymer white light-emitting devices and materials. Chem Soc Rev 38:3391–3400
2. Gong X, Ma WL, Ostrowski JC et al (2004) White electrophosphorescence from semiconducting polymer blends. Adv Mater 16:615–619
3. Gong X, Wang S, Moses D et al (2005) Multilayer polymer light-emitting diodes: white-light emission with high efficiency. Adv Mater 17:2053–2058
4. Zou J, Wu H, Lam C-S et al (2011) Simultaneous optimization of charge-carrier balance and luminous efficacy in highly efficient white polymer light-emitting devices. Adv Mater 23:2976–2980
5. Noh YY, Lee CL, Kim JJ et al (2003) Energy transfer and device performance in phosphorescent dye doped polymer light emitting diodes. J Chem Phys 118:2853–2864
6. Tu GL, Zhou QG, Cheng YX et al (2004) White electroluminescence from polyfluorene chemically doped with 1,8-napthalimide moieties. Appl Phys Lett 85:2172–2174
7. Liu J, Zhou QG, Cheng YX et al (2005) The first single polymer with simultaneous blue, green, and red emission for white electroluminescence. Adv Mater 17:2974–2978
8. Lee SK, Hwang DH, Jung BJ et al (2005) The fabrication and characterization of single-component polymeric white-light-emitting diodes. Adv Funct Mater 15:1647–1655
9. Jiang JX, Xu YH, Yang W et al (2006) High-efficiency white-light-emitting devices from a single polymer by mixing singlet and triplet emission. Adv Mater 18:1769–1773
10. Liu J, Chen L, Shao S et al (2007) Three-color white electroluminescence from a single polymer system with blue, green and red dopant units as individual emissive species and polyfluorene as individual polymer host. Adv Mater 19:4224–4228
11. Chien CH, Liao SF, Wu CH et al (2008) Electrophosphorescent polyfluorenes containing osmium complexes in the conjugated backbone. Adv Funct Mater 18:1430–1439
12. Liu J, Guo X, Bu L et al (2007) White electroluminescence from a single-polymer system with simultaneous two-color emission: polyfluorene as blue host and 2,1,3-benzothiadiazole derivatives as orange dopants on the side chain. Adv Funct Mater 17:1917–1925
13. Liu J, Shao S, Chen L et al (2007) White electroluminescence from a single polymer system: improved performance by means of enhanced efficiency and red-shifted luminescence of the blue-light-emitting species. Adv Mater 19:1859–1863
14. Poulsen DA, Kim BJ, Ma B et al (2010) Site isolation in phosphorescent bichromophoric block copolymers designed for white electroluminescence. Adv Mater 22:77–82
15. Shao S, Ding J, Wang L et al (2012) White electroluminescence from all-phosphorescent single polymers on a fluorinated poly(arylene ether phosphine oxide) backbone simultaneously grafted with blue and yellow phosphors. J Am Chem Soc 134:20290–20293
16. Adachi C, Kwong RC, Djurovich P et al (2001) Endothermic energy transfer: a mechanism for generating very efficient high-energy phosphorescent emission in organic materials. Appl Phys Lett 79:2082–2084

17. Ding J, Lue J, Cheng Y et al (2009) Effect of ancillary ligands on the properties of heteroleptic green iridium dendrimers functionalized with carbazole dendrons. J Organomet Chem 694:2700–2704
18. Wu F-I, Yang X-H, Neher D et al (2007) Efficient white-electrophosphorescent devices based on a single polyfluorene copolymer. Adv Funct Mater 17:1085–1092

Chapter 5
Spiro-Linked Hyperbranched Architecture for Electrophosphorescent Polymers

5.1 Background

Introducing phosphorescent emitters into polymers to harvest both singlet and triplet excitons in the EL process has been proved to be an effective approach to enhance the device efficiency [1–6]. To ensure that all the triplet excitons are confined effectively on the phosphorescent emitters, triplet energy back transfer (TEBT) from phosphors to polymer hosts should be inhibited (see Sect. 1.3 in Chap. 1). Basically, there are two ways to prevent the TEBT process in PhPLEDs:

1. Elevating E_Ts of the polymer host. E_Ts of polymer hosts can be elevated by reducing the conjugation extent through *meta*-linkage [7–9], inserting saturated atoms in the backbone [10, 11], and so on. However, for most polymer hosts, the E_Ts are lower than 2.70 eV, which are not suitable for blue dopants. Developing efficient polymer hosts with E_Ts higher than 2.70 eV has been proved to be a big challenge according to the previous reports (see Sect. 1.4 in Chap. 1).
2. Separating the dopants from hosts [3, 12, 13]. As the Dexter energy transfer rate is strongly dependent on the distance between the energy donor and acceptor, separating the dopants from hosts to increase their distance is expected to inhibit the TEBT process. For example, A. B. Holmes demonstrated that [12] by introducing a long alkyl spacer between the phosphor and the polymer main-chain, the intramolecular TEBT process could be weakened due to the enhancement of the distance between their triplet centers. As a result, the EQE increased from 1.1 to 2.0 % as compared with that without spacer. However, the intermolecular TEBT can still take place from the phosphor at one polymer chain to another neighboring polymer chain, resulting in the loss of triplet excitons.

In view of the limitation of the known methods in inhibiting TEBT process, it is of great significance to provide new approaches to solve this problem.

S. Shao, *Electrophosphorescent Polymers Based on Polyarylether Hosts*,
Springer Theses, DOI 10.1007/978-3-662-44376-7_5

5.2 Molecular Design

In this thesis, we proposed a spiro-linked hyperbranched architecture for PhPs, to modulate the TEBT process [14]. As we know, the TEBT process (which is a Dexter energy transfer process) is dominated via electron exchange that requires a good intermolecular overlap between the pertinent molecular orbitals of the donor and acceptor [15]. Therefore, we propose that a spiro linkage between two π conjugated halves can render a perpendicular arrangement between them [16] and thus reduce their electronic interactions to a large extent. With this linkage, the intramolecular TEBT is inhibited. In addition, the hyperbranched architecture makes sure that the Ir complex core is isolated from neighboring polymer chains, so the intermolecular TEBT is also prevented. Combining these two points, we believe the spiro-linked hyperbranched concept can inhibit both the intra- and intermolecular TEBT process of the PhPs (Fig. 5.1).

To support our hypothesis, a prototype oligomer named Sp–Ir–TF was designed (Fig. 5.2), in which the guest tris[9,9-dioctyl-2-(pyridinyl-2')fluorene]iridium(III) ($Ir(FP)_3$) [17, 18] is encapsulated by the host trifluorene (TF) via the spiro bridge. Here the oligomer Sp–Ir–TF, rather than the polymer analog, was synthesized for several reasons. Firstly, unlike the polymer, oligomer Sp–Ir–TF is easy to prepare and purify without batch-to-batch variations. Secondly, Sp–Ir–TF displays monodispersed and well-defined molecular structure, which leads to the direct correlation between the chemical structures and physical properties. Finally, similar vertical conformation exists between the Ir-complex core with the trifluorene host in Sp–Ir–TF as in the polymer, which is supposed to account for the constraint of TEBT process.

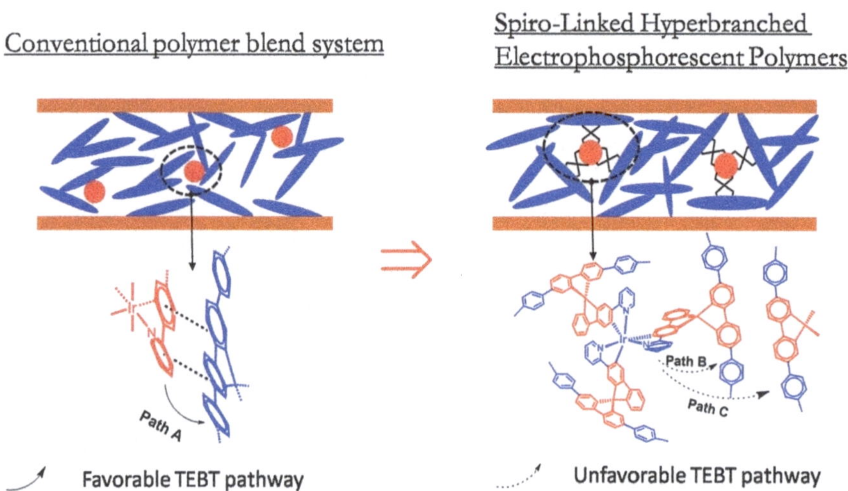

Fig. 5.1 Design strategy of the spiro-linked hyperbranched PhPs

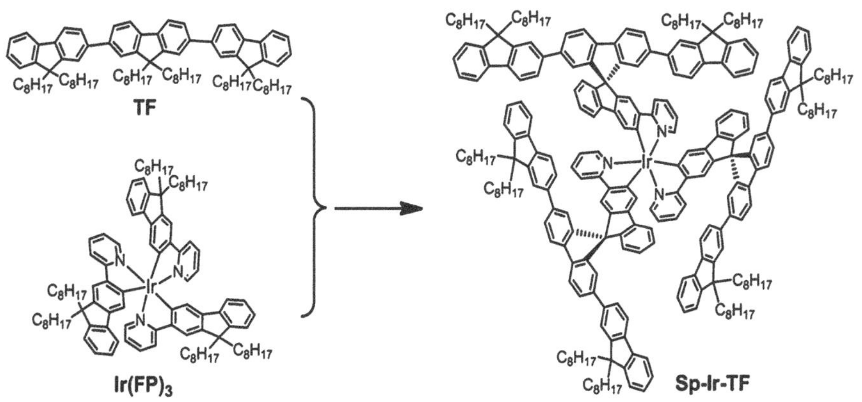

Fig. 5.2 Chemical structures of the polymer blend and spiro-linked hyperbranched PhPs. Reproduced from Ref. [14] by permission of John Wiley and Sons Ltd

It should be noted that the E_T of Ir(FP)$_3$ is estimated to be ~2.3 eV, which is somewhat higher than that of TF (~2.2 eV) [19]. So exothermic TEBT is prone to occur spontaneously from Ir(FP)$_3$ to TF when they are simply mixed together. This enables us to evaluate whether the spiro-linked hyperbranched architecture in Sp–Ir–TF can prevent the process of TEBT as expected.

5.3 Results and Discussions

Synthesis and characterization As shown in Scheme 5.1, TF was synthesized by Suzuki coupling of **1** and **2** in a high yield. Reaction between **3** and $IrCl_3$–$3H_2O$ in $CH_3CH_2OCH_2CH_2OH$/H2O (3/1, v/v) provided the dimer **4** in nearly quantitative yield. The facial complex Ir(FP)$_3$ was obtained by a low-temperature reaction of ligand **3** and **4** with CF_3SO_3Ag as the promoter and *o*-dibromobenzene as the solvent [20]. Similarly, the key intermediate **11** for the synthesis of Sp–Ir–TF was obtained from dimer **10** and ligand **9** which was derived from the reaction between 2,7-dibromofluoreone with the corresponding arly lithuium of **8**. Finally, Suzuki coupling of **11** and **1** gave the desired product Sp–Ir–TF. All the structures of the intermediates were confirmed by ^{1}H NMR spectra. Additional measurements, including ^{13}C NMR, MALDI-TOF MS, and elementary analysis (EA) were also performed for TF, Ir(FP)$_3$ and Sp–Ir–TF for further confirmation. Especially, the ^{1}H NMR and MALDI-TOF MS of Sp–Ir–TF have been shown in Figs. 5.3 and 5.4, respectively.

The UV-Vis spectrum of Sp–Ir–TF in dichloromethane is presented in Fig. 5.5a. The weak absorption band in the range of 400–510 nm is assigned to the metal-to-ligand charge transfer (MLCT) transition of the Ir complex, and the high-energy absorption peaks at 324/338/356 nm are attributed to the π-π* transitions of the

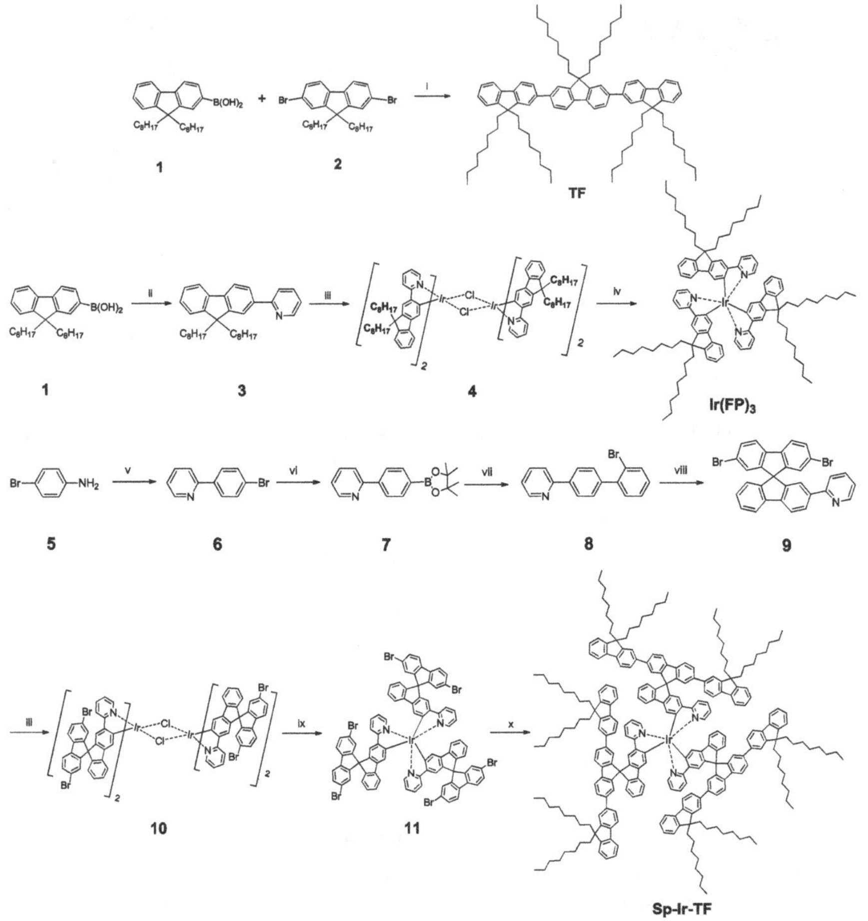

Scheme 5.1 Synthetic routes of the compounds. Synthetic conditions: i. $Pd(PPh_3)_4$, K_2CO_3, toluene/H_2O, Aliquat 336, 90 °C; ii. 2-bromopyridine, $Pd(PPh_3)_4$, K_2CO_3, toluene/H_2O, Alquat 336, 90 °C; iii. $IrCl_3$–$3H_2O$, $CH_3CH_2OCH_2CH_2OH$, reflux; iv. (1) CF_3SO_3Ag, CH_3CN, (2) **3**, o-dichlorobenzene, 110 °C; v. (1) $NaNO_2$, HCl, (2) pyridine, Na_2CO_3; vi. Bis(pinacolato) diboron, $PdCl_2$(dppf)-CH_2Cl_2, KOAc, DMF, 80 °C; vii. *o*-dibromobenzene, $Pd(PPh_3)_4$, K_2CO_3, toluene/H_2O, Aliquat 336, 90 °C; viii. (1) Mg, THF, 60 °C, (2) 2,7-dibromofluorenone, (3) BF_3–Et_2O; ix. (1) CF_3SO_3Ag, CH_3CN, (2) **9**, *o*-dichlorobenzene, 110 °C; x. **1**, $Pd(PPh_3)_4$, K_2CO_3, toluene/H_2O, Aliquat 336, 90 °C

C^N ligands and TF fragments, respectively. Noticeably, the absorption spectrum of Sp–Ir–TF is in excellent accordance with that calculated from TF and $Ir(FP)_3$ (i.e. $3\varepsilon_{TF} + \varepsilon_{Ir(FP)3}$), and no new emission corresponding to the electron communication between TF and $Ir(FP)_3$ appears. This observation unambiguously indicates that the spiro linkage between TF and $Ir(FP)_3$ makes their electron coupling negligible [21]. In addition, as shown in Fig. 5.5b, the emission peak of Sp–Ir–TF is blue-shifted by

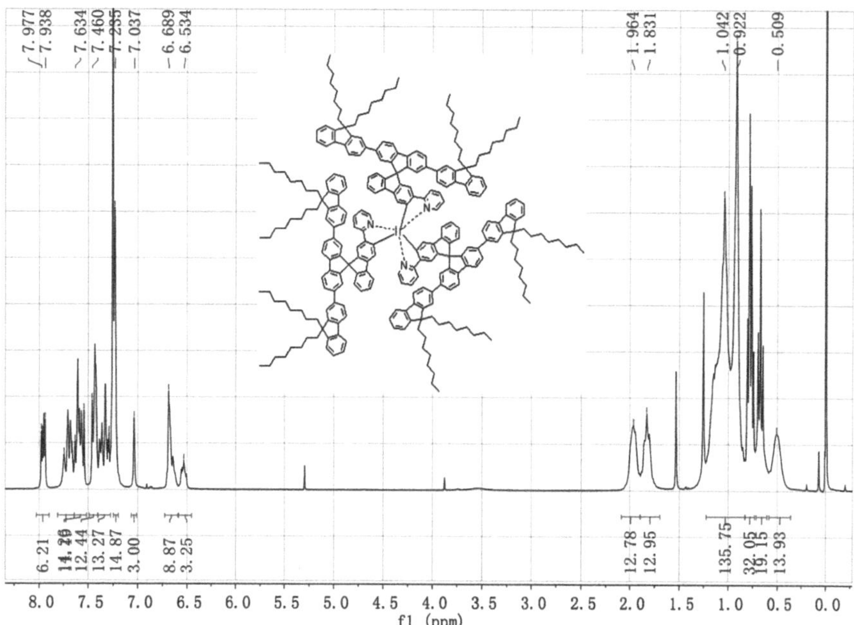

Fig. 5.3 ^{1}H NMR spectrum of Sp–Ir–TF

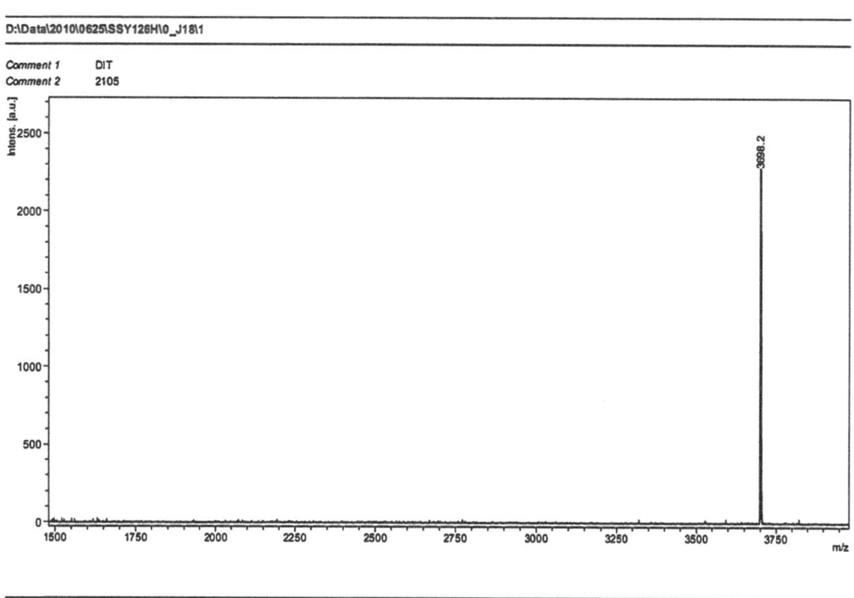

Fig. 5.4 MALDI-TOF MS spectrum of Sp–Ir–TF

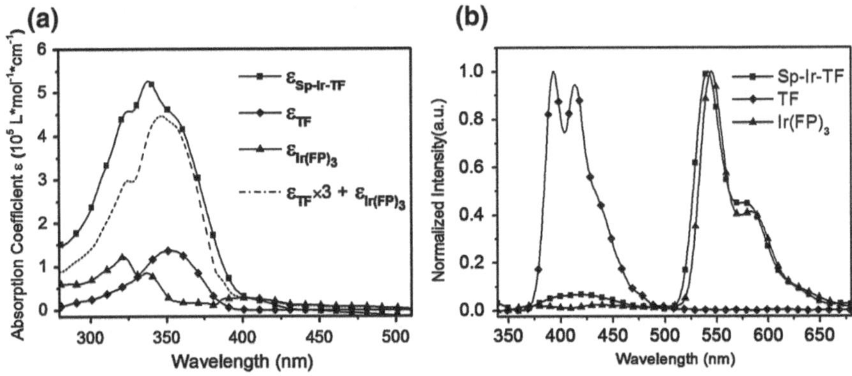

Fig. 5.5 **a** The UV-Vis absorption spectra of TF, $Ir(FP)_3$, and Sp–Ir–TF in dilute dichloromethane solutions. For comparison, the absorption spectrum of Sp–Ir–TF is also estimated according to the equation: $\varepsilon_{Sp\text{–}Ir\text{–}TF} = 3\varepsilon_{TF} + \varepsilon_{Ir(FP)3}$. **b** The PL spectra of TF, $Ir(FP)_3$, and Sp–Ir–TF in dilute toluene solutions. Reproduced from Ref. [14] by permission of John Wiley and Sons Ltd

about 3 nm in comparison to $Ir(FP)_3$, indicative of the existence of the shielding effect of the Ir complex core by the peripheral TF fragments in Sp–Ir–TF. Furthermore, the emission from TF is nearly indiscernible in the PL spectrum of SP–Ir–TF, suggesting the efficient singlet energy transfer from TF to Ir complex core via Förster process.

The PL quantum yields (PLQYs) of the compounds in toluene were measured with the excitation wavelength of 410 nm. Under this excitation wavelength, the absorption of TF can be ignored. Consequently, the Förster energy transfer from the TF fragments to the Ir complex core is constrained, and the PLQY of Sp–Ir–TF is solely determined by the intramolecular TEBT. As listed in Table 5.1, Sp–Ir–TF gives a PLQY of 0.26, very close to that of $Ir(FP)_3$ (0.23). The comparable PLQYs between Sp–Ir–TF and $Ir(FP)_3$ imply that the spiro-bridged TF units are not the phosphorescence quencher for $Ir(FP)_3$ any longer in Sp–Ir–TF. That is to say, the intramolecular TEBT is completely suppressed, otherwise, Sp–Ir–TF would exhibit

Table 5.1 Photophysical and electrochemical properties of TF, $Ir(FP)_3$ and Sp–Ir–TF

Compound	Absλ $_{max}^{a}$(nm)	PLλ $_{max}^{b}$(nm)	PLQY[c]	E_g^d (eV)	E_T^e (eV)	$E_{onset,ox}^{f}$ (V)	HOMO[g] (eV)	LUMO[g] (eV)
TF	354	394, 414	–	3.20	–	0.83	−5.63	−2.43
$Ir(FP)_3$	321, 338	545, 587	0.23	2.44	2.3	0.16	−4.96	−2.52
Sp–Ir–TF	324, 338, 356	542, 579	0.26	2.44	2.3	0.22	−5.02	−2.58

[a] Measured in CH_2Cl_2

[b] Measured in toluene at 298 K

[c] PL quantum efficiency tested in toluene using *fac*-tris(2-phenylpyridine)iridium (*fac-Ir(ppy)₃*, PLQY: 0.40) as a standard

[d] Energy gap calculated from the onset of absorption

[e] Triplet energy calculated from the first vibration of the phosphorescent spectrum at 77 K

[f] Onset of the first oxidation curve (vs. Fc^+ /Fc)

[g] The HOMO and LUMO energy levels were calculated according to the equations: E_{HOMO} = −e [$E_{onset,\ ox}$ + 4.8 V], and $E_{LUMO} = E_g + E_{HOMO}$

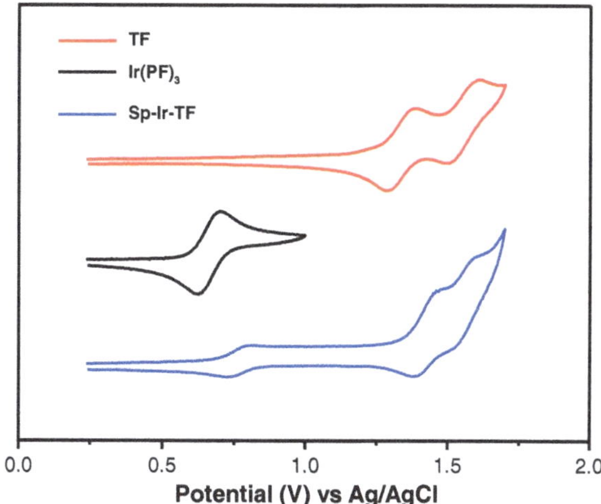

Fig. 5.6 Cyclic voltammetry curves of TF, Ir(FP)$_3$, and Sp–Ir–TF

lower PLQY than Ir(FP)$_3$. Figure 5.6 shows the oxidization curves of TF, Ir (FP)3, and Sp–Ir–TF in dichloromethane. Sp–Ir–TF shows two sets of oxidation peaks with the onsets at 0.68 and 1.35 eV, respectively, which are very close to those of Ir (FP)$_3$ and TF. This observation again indicates that the Ir-complex core and the peripheral TF units are electronically independent in Sp–Ir–TF.

These exciting observations promoted us to further study the PL and EL properties of Sp–Ir–TF. To avoid triplet–triplet (T–T) annihilation, Sp–Ir–TF was dispersed in TF to prepare films with doping concentrations of 0.2, and 2 mol% [22], respectively. For comparison, Ir(FP)$_3$-based control films with the same Ir-content were also prepared. Figure 5.7 depicts the PL and EL spectra of these two kinds of films. Obviously, the relative intensity of the yellow emission to the blue emission is higher in the Sp–Ir–TF/TF films than those in the Ir(FP)$_3$/TF ones, indicating that the loss of excitons of Ir complex is inhibited by the spiro-linked hyperbranched architecture.

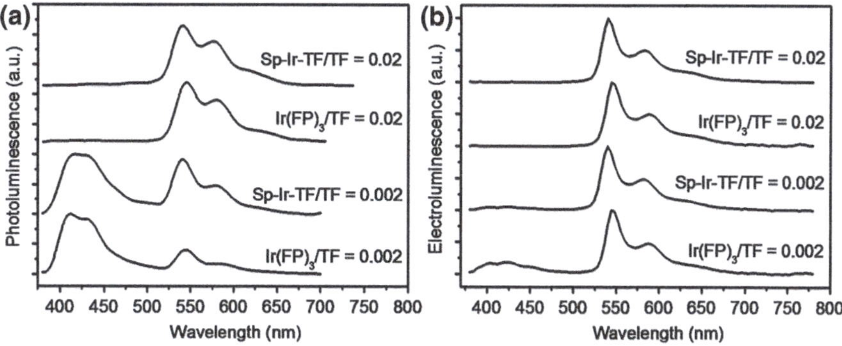

Fig. 5.7 The PL (excited at 365 nm) (**a**) and EL spectra (**b**) of Sp–Ir–TF/TF thin films compared with Ir(FP)$_3$/TF ones. Reproduced from Ref. [14] by permission of John Wiley and Sons Ltd

In addition, in the EL spectra, only the emission from the triplet state of Ir complex is observed with little or no singlet emission from TF. This is quite different from the PL spectra especially at 0.2 mol% doping concentration, where considerable singlet emission can be observed. Moreover, the HOMO and LUMO energy levels of $Ir(FP)_3$ (HOMO −5.0 eV and LUMO −2.5 eV) fall within those of TF (HOMO −5.6 eV and LUMO −2.4 eV). Therefore, we conclude that charge trapping rather than energy transfer is the predominant mechanism in the EL process [22, 23].

The EQE versus current density curves of these two kind of films, Sp–Ir–TF/TF and $Ir(FP)_3$/TF, are shown in Fig. 5.8. Obviously, Sp–Ir–TF/TF exhibits much higher EQEs than $Ir(FP)_3$/TF. For example, the EQEs at a current density of 2 mA/cm^2 for Sp–Ir–TF/TF = 0.002 and 0.02 are 1.0 and 1.5 %, respectively, which are about 5–8 times those of $Ir(FP)_3$/TF. Correspondingly, the luminescence significantly increases from 7–18 cd/m^2 of $Ir(FP)_3$/TF to 68–104 cd/m^2 of Sp–Ir–TF/TF (Table 5.2). Given that the same number of triplet excitons should be created in Sp–Ir–TF/TF as in $Ir(FP)_3$/TF due to charge trapping on Ir complex under electrical excitation, the improvement of the device performance can be rationalized by the molecular structure difference between $Ir(FP)_3$ and Sp–Ir–TF.

Since Dexter transfer is a short-range (<10 Å) process, only the TF molecules near the Ir complex can become the phosphorescence quenching sites. In $Ir(FP)_3$/TF films, the adjacent TF can readily adopt appropriate conformation with respect to the Ir complex to facilitate efficient TEBT, leading to the triplet exciton deactivation. In contrast, the unique encapsulated structure in Sp–Ir–TF/TF system makes sure that the periphery around Ir complex is almost full of spiro-bridged TF units. And then, the negligible electron interactions between them, as demonstrated above, prevent the TEBT process from $Ir(FP)_3$ to the spiro-bonded TFs. As a consequence, triplet excitons can be effectively confined on the Ir complex to realize high EL efficiency.

To verify this effect further, we recorded the transient PL of Sp–Ir–TF/TF and Ir $(FP)_3$/TF thin films. The excited wavelength is the same as that used for the

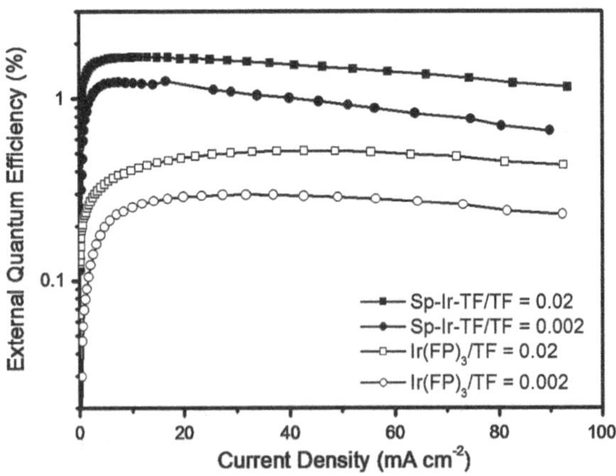

Fig. 5.8 EQE-*J* curves of the devices. Reproduced from Ref. [14] by permission of John Wiley and Sons Ltd

Table 5.2 Device performance and lifetime data of the films under investigation. Reproduced from Ref. [14] by permission of John Wiley and Sons Ltd

Films	EQE[a] (%)	L^a (cd m^{-2})	EQE[b] (%)	L^b (cd m^{-2})	τ_1^c (μs)	A_1^c	τ_2 (μs)	A_2
Ir(FP)3/ TF = 0.02	0.27	18	0.47	311	0.50	0.77	5.04	0.23
Sp–Ir–TF/ TF = 0.02	1.5	104	1.7	1170	0.80	0.68	9.45	0.32
$Ir(FP)_3$/ TF = 0.002	0.12	7	0.29	165	0.35	0.52	1.84	0.48
Sp–Ir–TF/ TF = 0.002	1.0	68	1.1	806	0.68	0.68	3.09	0.32

[a] Values at 2 mA cm^{-2}

[b] Values at 20 mA cm^{-2} . [c] τ and A are the photoluminescence lifetimes and relative intensity contributions, respectively, according to the biexponential decay equation: $I_{PL}(t) = A_1\exp(t/\tau_1) + A_2\exp(t/\tau_2) + c$, where I_{PL} is the photoluminescence intensity and c is a constant

measurement of PLQYs, which allows direct excitation of the Ir complex. Figure 5.9 shows the decay curves of the PL intensity at room temperature and the biexponential fit lines. It is noteworthy that the PL decays slower in Sp–Ir–TF/TF than in the $Ir(FP)_3$/TF, which explicitly proves that the TEBT is effectively prohibited in Sp–Ir–TF.

One may think that, the efficiency enhancement of Sp–Ir–TF/TF system can be attributed to the inhibited triplet–triplet annihilation of the triplet excitons induced by the capsulating effect of the peripheral TF. We propose that this possibility can be ruled out because (1) in both Sp–Ir–TF/TF and $Ir(FP)_3$/TF system, the Ir-loadings are very low. In general, triplet–triplet annihilation occurs only when the Ir-concentration is higher than 2 mol% [22, 23]. (2) the luminous efficiency increases as the Ir-content increases, indicating that the triplet–triplet annihilation effect does not take the dominant place in this case.

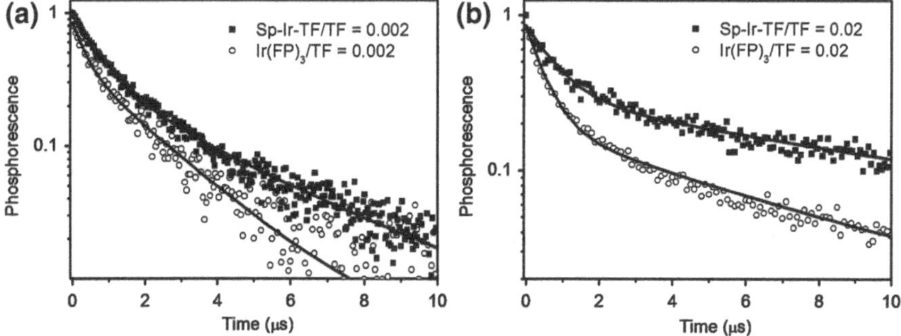

Fig. 5.9 PL intensity decay curves of films of $Ir(FP)_3$/TF = 0.002 and Sp–Ir–TF/TF = 0.002 (**a**), and $Ir(FP)_3$/TF = 0.02 and Sp–Ir–TF/TF = 0.02 (**b**) (Excited at 410 nm, monitored at 545 nm). Reproduced from Ref. [14] by permission of John Wiley and Sons Ltd

In summary, by adopting a spiro-linked hyperbranched architecture in Sp–Ir–TF, an about 5–8-folds improvement of the EQEs as well as longer lifetimes has been achieved in comparison to the conventional blending system. This unique architecture is proposed to be able to prevent both the intra- and inter-molecular TEBT from phosphor guest to conjugated polymer host in PhPs, and thereby achieve enhanced device performance. This strategy provides us a new approach to modulating the TEBT process and the molecular design principles can be generalized for the de nova design of high-performance PhPs.

5.4 Experimental Section

General Information All chemicals and reagents were used as received from commercial sources without further purification. Solvents for chemical synthesis were purified according to the standard procedures. All air and water sensitive reactions were performed under argon. ^{1}H NMR spectra were recorded with a Bruker Avance 300 NMR spectrometer. Elemental analysis was performed using a Bio-Rad elemental analysis system. MALDI-TOF mass spectra were performed on a AXIMA CFR MS apparatus (COMPACT). UV–Visible (UV–Vis) absorption and photoluminescence spectra were measured with a Perkin-Elmer Lambda 35 UV–Vis spectrometer and a Perkin-Elmer LS 50B spectrofluorometer, respectively. Phosphorescence spectra at 77 K were measured in a toluene/ethanol/methanol (5:4:1) mixed solvent. The triplet energies were estimated as the maximum of the first vibronic mode ($S_0^{v=0} \leftarrow T_1^{v=0}$) of the corresponding phosphorescence spectra at 77 K. The PL decay curves were obtained from a Lecroy Wave Runner 6,100 digital oscilloscope (1 GHz) by using a tunable laser (pulse width = 4 ns, gate = 50 ns) as excitation source (Continuum Sunlite OPO). CV experiments were performed on an EG&G 283 (Princeton Applied Research) potentiostat/galvanostat system. The scan rate was 100 mV/s. Solutions of the samples were tested in CH_2Cl_2 using ferrocene as an internal reference and n-Bu_4NClO_4 as the supporting electrolyte. The HOMO energy levels were calculated according to the equation $E_{HOMO} = -e\ [E_{onset,\ ox} + 4.8\ V]$, in which $E_{onset,\ ox}$ was the onset of the first oxidation potential. The LUMO energy levels were calculated according to the equation $E_{LUMO} = E_g + E_{HOMO}$, where E_g is the optical band gap estimated from the onset of the absorption spectrum.

Device Fabrication and Characterization OLEDs were fabricated with structures of ITO/PEDOT: PSS (40 nm)/emissive layer (40 nm)/TPBI (50 nm)/LiF (1 nm)/Al (200 nm). Indium-tin oxide (ITO) with a sheet resistance 10–20 Ω/square was used as the substrate and it was cleaned with surfactant and then deionized water. Oxygen plasma treatment was made for 25 min to improve the contact angle before film coating. Poly(3,4-ethylenedioxythiophene):Poly(styrenesulfonate) (PEDOT:PSS) film was spin-coated with thickness of 40 nm onto the ITO glass to improve the hole injection and to avoid the possibility of current leakage. PEDOT: PSS film was baked at 120 °C in air for 40 min. The solution of all materials in toluene was spin-coated on top of PEDOT:PSS film. Typical thickness of the emitting layer was 40 nm. Successively, a 50-nm-thick film of 1,3,5-tris(2-N-phenylbenzimidazolyl)benzene (TPBI) was thermally evaporated on top of the EML at a base pressure less than 10^{-6} Torr (1 Torr=133.32 Pa) to serve as a exciton-blocking layer. Finally, a 1-nm-thick film of LiF, and a 200-nm-thick film of Al were thermally evaporated as an electron injection cathode. The deposition speed and the thickness of TPBI, LiF, and Al were monitored by a thickness/rate meter. The cathode area defines the active area of the device. The typical active area of the devices in this study is 0.14 cm^2. The EL spectra were measured using a PR650 spectra colorimeter. The current-voltage and brightness-voltage curves of devices were measured using a Keithley 2,400/2,000 source meter and a calibrated silicon photodiode. All the device performance tests were carried out at room temperature under ambient conditions.

Synthesis **TF**. Compound **1** (1.04 g, 2.4 mmol), **2** (0.55 g, 1 mmol), $Pd(PPh_3)_4$ (0.023 g, 0.02 mmol), aqueous K_2CO_3 (2 M, 10 mL), and toluene (20 mL) were mixed in a flask. The mixture was degassed and then refluxed for 24 h under an argon atmosphere. After being cooled, the organic layer was separated and washed with brine and water. After evaporation of the solvent, the residue was purified via chromatography over silica gel (petroleum ether: dichloromethane = 10:1) to give a crude product. Recrystallizing the crude product from ethanol for three times gave the pure product as a colorless crystal (0.82 g,) in a yield of 70 %. ^{1}H NMR (300 MHz, $CDCl_3$, ppm) δ: 7.81 (d, J = 7.9 Hz, 2H), 7.79 (d, J = 7.9 Hz, 2H), 7.74 (d, J = 7.4 Hz, 2H), 7.68–7.65 (m, 4H), 7.64 (s, 2H), 7.62 (s, 2H), 7.39–7.29 (m, 6H), 2.12–2.01 (m, 12H), 1.18–1.08 (m, 60H), 0.83–0.77 (m, 18H), 0.75–0.71 (m, 12H). MALDI-TOF MS: Calcd for $C_{87}H_{122}$: 1166.9. Found: 1166.9 (M^+). Anal. Calcd for $C_{87}H_{122}$: C, 89.47; H, 10.53. Found: C, 89.12; H, 10.45.

3. A similar procedure as the synthesis of **TF** was performed from **1** (0.87 g, 2 mmol) and 2-bromopyridine (0.24 mL, 2.5 mmol) to afford **3** (0.89 g) in a yield of 95 %.^{1}H NMR (300 MHz, $CDCl_3$, ppm) δ: 8.74 (d, 1H, J = 4.6 Hz), 8.01 (s, 1H), 8.00 (d, 1H, J = 4.6 Hz), 7.82–7.73 (m, 4 H), 7.33–7.31 (m, 3H), 7.29–7.24 (m, 1H), 2.07–1.99 (m, 4H), 1.18–1.03 (m, 20H), 0.79 (t, 6H, J = 6.9 Hz), 0.67–0.60 (m, 4H).

4. A mixture of **3** (3.2 g, 6.84 mmol), $IrCl_3$-$3H_2O$ (1.15 g, 3.26 mmol), 2-ethoxyethanol (30 mL) and water (10 mL) was refluxed under argon for 24 h. After being cooled to room temperature, the precipitate was collected by filtration and washed with water and ethanol. Then the crude product was purified by column chromatography on silica gel with petroleum ether: ethyl acetate (10:1) as eluent to give the chloro-bridged iridium dimer as a brown powder (2.43 g, 64.3 %).^{1}H NMR (300 MHz, $CDCl_3$, ppm) δ 9.36 (d, J = 5.1 Hz, 4H), 7.96 (d, J = 8.0 Hz, 4H), 7.83 (t, J = 7.7 Hz, 4H), 7.44 (s, 4H), 7.19–7.00 (m, 16H), 6.89 (t, J = 6.5 Hz, 4H), 6.35 (s, 4H), 1.88–1.74 (m, 16H), 1.23–0.82 (m, 80H), 0.75 (t, J = 5.5 Hz, 12H), 0.71 (t, J = 5.5 Hz, 12H), 0.55 (s, 16H).

$Ir(FP)_3$. $Ir(FP)_3$ was obtained as a facial complex according to a selective low-temperature synthetic route for facial tris-cyclometalated iridium(III) complexes (*Inorganic Chemistry* **2007**, *46*, 7800). Compound **4** (0.15 g, 0.065 mmol) in 40 mL of acetonitrile was heated to dissolve all of the chloro-bridged dimer. A 10 mL acetonitrile solution of $AgOSO_2CF_3$ (0.037 g, 0.142 mmol) was added to the Ir solution. This mixture was heated in the dark at 60 °C for 2 h and a gray precipitate appeared. The reaction mixture was filtered to separate a yellow solution from the gray AgCl precipitate. The solution was concentrated in vacuo to give a brown powder. Successively, a mixture of the brown powder and **3** (0.064 g, 0.14 mmol) was heated in 10 mL *o*-dichlorobenzene at 110 °C for 72 h. After evaporation of the solvent, the residue was purified via chromatography over silica gel (petroleum ether: dichloromethane = 5:1) to give the pure product (0.045 g, 22 %). ^{1}H NMR (300 MHz, $CDCl_3$, ppm) δ: 7.93 (d, J = 8.2 Hz, 3H), 7.80 (s, 3H), 7.58 (t, J = 7.3 Hz, 3H), 7.37 (s, 3H), 7.23 (m, 6H), 7.11 (d, J = 7.5 Hz, 3H), 7.07 (t, J = 7.5 Hz, 3H), 6.96 (t, J = 7.5 Hz, 3H), 6.55 (s, 3H), 2.17–2.03 (m, 6H), 2.00–1.86 (m, 6H), 1.16–1.04 (m, 60H), 0.83–0.76 (m, 18H), 0.72–0.58 (m, 12H). MALDI-TOF MS: Calcd for $C_{102}H_{132}IrN_3$: 1592.0. Found: 1592.0 (M^+). Anal. Calcd for $C_{102}H_{132}IrN_3$: C, 76.93; H, 8.36; N, 2.64. Found: C, 76.71; H, 8.30; N, 2.59.

6. An aqueous solution of $NaNO_2$ (8 M in water, 45 mL, 0.36 mmol) was slowly added to a suspension of *p*-bromoaniline (31 g, 0.18 mmol) in concentrated HCl (60 mL) at 0 °C. The mixture was stirred for 1 h at 0 °C and was slowly poured into pyridine (750 mL). The resulting brown solution was stirred at 40 °C for 4 h and then Na_2CO_3 (200 g) was added and the slurry was stirred for a further 18 h. Excess pyridine was then removed by evaporation and the organic product was extracted with dichloromethane. The organic fractions were dried over Na_2SO_4 and evaporated to dryness. Column chromatography over silica gel (petroleum ether: ethyl acetate = 4:1) yielded the pure product (14.7 g, 35 %). ^{1}H NMR (300 MHz, $CDCl_3$, ppm) δ: 8.69 (d, J = 4.3 Hz, 1H), 7.88 (d, J = 8.4 Hz, 2H), 7.80–7.75 (m, 1H), 7.71 (d, J = 7.8 Hz, 1H), 7.60 (d, J = 8.4 Hz, 2H), 7.29–7.24 (m, 1H).

7. Into an argon flushed flask was charged with **6** (2.18 g, 9.3 mmol), bis (pinacolato)diboron, 2.6 g (10.2 mmol), potassium acetate (2.73 g, 27.9 mmol) and 0.23 g (0.28 mmol) of [1,1'-bis(diphenylphosphino)ferrocene]dichloropalladium–dichloromethane. 20 mL dimethyl sulfide was then added, and the mixture was heated at 80 °C for 24 h. The reaction mixture was cooled down and extracted with diethyl ether (50 mL × 3). The combined organic layer was dried over Na_2SO_4. After the solvent was removed, the residue was run through a silica gel column with petroleum ether: ethyl acetate (4:1) as eluent to give 2.1 g white powder (yield, 81%). ^{1}H NMR (300 MHz, $CDCl_3$, ppm) δ: 8.72 (d, J = 4.8 Hz, 1H), 8.01 (d, J = 8.2 Hz, 2H), 7.92 (d, J = 8.2 Hz, 2H), 7.80–7.71 (m, 2H), 7.27–7.23 (m, 1H), 1.36 (s, 12H).

8. **8** was obtained as a light yellow powder (1.5 g, 65 %) from **7** (2.1 g, 7.47 mmol) and *o*-dibromobenzene (2.7 mL, 22.4 mmol) according to a procedure similar to that of **TF.**^{1}H NMR (300 MHz, $CDCl_3$, ppm) δ: 8.73 (d, J = 4.7 Hz, 1 H), 8.08 (d, J = 8.2 Hz, 2 H), 7.81–7.79 (m, 2 H), 7.69 (d, J = 8.1 Hz, 1 H), 7.54 (d, J = 8.3 Hz, 2 H), 7.38 (d, J = 4.2 Hz, 2 H), 7.29–7.19 (m, 2 H).

9. To a well-degassed solution of **8** (0.31 g, 1 mmol) in THF (20 mL) was added *n*-BuLi (0.4 mL, 1 mmol, 2.5 M in hexane) at −78 °C. The mixture was stirred for 30 min, and then a solution of 2, 7-dibromofluorone (0.37 g, 1.1 mmol) in 5 mL tetrahydrofuran was added. After stirring at room temperature for 3 h, the reaction mixture was quenched with aqueous solution of NH_4Cl and extracted with dichloromethane. The organic layer was washed with brine and water. After removal of the solvent, the residue was purified via chromatography over silica gel (petroleum ether: ethyl acetate = 15:2) to give the intermediate alcohol as a slightly yellow powder. ^{1}H NMR (300 MHz, $CDCl_3$, ppm) δ: 8.65 (d, J = 4.8, 1H), 8.42 (dd, J = 8.0, 0.9 Hz, 1H), 7.76 (td, J = 7.8, 1.7 Hz, 1H), 7.62–7.51 (m, 2H), 7.39 (dd, J = 7.5, 1.2 Hz, 1H), 7.35–7.29 (m, 6H), 7.24–7.18 (m, 1H), 6.97 (dd, J = 7.5, 1.5 Hz, 1H), 6.96 (d, J = 7.8, 2H), 6.20 (d, J = 8.3 Hz, 2H), 2.40 (s, 1H). This powder was dissolved in dichloromethane and BF_3–Et_2O (0.14 mL, 1.1 mmol) was added. After stirring at room temperature for 8 h, the mixture was washed with water, and the organic layer was dried over Na_2SO_4. After evaporation of the solvent, the residue was purified over silica gel column with petroleum ether: dichloromethane = 1:1 as the eluent. A white powder (0.41 g) was obtained in a yield of 75 %. ^{1}H NMR (300 MHz, $CDCl_3$, ppm) δ: 8.61 (d, J = 4.4 Hz, 1H), 8.11 (d, J = 7.6 Hz, 1H), 7.97 (d, J = 8.0 Hz, 1H), 7.89 (d, J = 8.0 Hz, 1H), 7.68 (d, J = 8.0 Hz, 3H), 7.62 (d, J = 8.0 Hz, 1H), 7.49 (dd, J = 8.0, 2.0 Hz, 2H), 7.43 (td, J = 7.6, 1.6 Hz, 1H), 7.38 (d, J = 1.6 Hz, 1H), 7.19–7.15 (m, 2H), 6.87 (d, J = 1.6 Hz, 2H), 6.72 (d, J = 7.6 Hz, 1H).

10. **10** was obtained as a yellow powder (0.32 g, 94 %) from **9** (0.3 g, 0.54 mmol) and $IrCl_3$–$3H_2O$ (0.091 g, 0.26 mmol) according to a procedure similar to that of **4.** ^{1}H NMR (300 MHz, $CDCl_3$, ppm) δ: 9.34 (d, J = 5.3 Hz, 1H), 7.68 (d, J = 8.1 Hz, 1H), 7.62 (dd, J = 8.2, 1.5 Hz, 2H), 7.45–7.40 (m, 2H), 7.23–7.09 (m, 3H), 6.94 (td, 7.5, 0.9 Hz, 1H), 6.89–6.84 (m, 2H), 6.75 (d, J = 1.7 Hz, 1H), 6.74 (d, J = 1.7 Hz, 1H), 6.51 (d, J = 7.5 Hz, 1H), 6.48 (s, 1H).

11. **11** was obtained as a yellow powder (0.083 g, 30 %) from **10** (0.20 g, 0.076 mmol) and **9** (0.088 g, 0.16 mmol) according to a procedure similar to that of **Ir(FP)$_3$**. [1] ^{1}H NMR (300 MHz, $CDCl_3$, ppm) δ: 7.70 (d, J = 3.4 Hz, 1H), 7.68–7.64 (m, 2H), 7.56–7.44 (m, 5H), 7.39–7.33 (m, 2H), 7.10–7.08 (m, 2H), 7.03 (t, J = 7.5 Hz, 1H), 6.82 (d, J = 1.7 Hz, 1H), 6.85–6.78 (m, 1H), 6.64 (d, J = 7.5 Hz, 1H).

Sp–Ir–TF. The synthetic procedure was similar to that of Ir(FP)$_3$. A yellow powder (0.10 g) was obtained from **11** (0.062 g, 0.034 mmol) and **1** (0.144 g, 0.31 mmol) in a yield of 80 %. ^{1}H NMR (300 MHz, $CDCl_3$, ppm) δ: 7.97 (d, J = 3.6 Hz, 3H), 7.94 (d, J = 3.9 Hz, 3H), 7.74–7.66 (m, 12H), 7.63–7.54 (m, 15H), 7.46–7.43 (m, 12H), 7.39–7.28 (m, 15H), 7.24–7.20 (m, 15H), 7.04 (s, 3H), 6.69–6.59 (m, 9H), 6.56–6.51 (m, 3H), 2.07–1.91 (m, 12H), 1.85–1.80 (m, 12H), 1.15–0.92 (m, 120H), 0.81–0.75 (m, 30H), 0.67 (s, 18H), 0.57–0.40 (m, 12H). MALDI-TOF MS: Calcd for $C_{264}H_{294}IrN_3$: 3698.2. Found: 3698.3 (M^+). Anal. Calcd for $C_{264}H_{294}IrN_3$: C, 85.67; H, 8.01; N, 1.14. Found: C, 85.43; H, 7.96; N, 1.05.

References

1. Chen XW, Liao JL, Liang YM et al (2003) High-efficiency red-light emission from polyfluorenes grafted with cyclometalated iridium complexes and charge transport moiety. J Am Chem Soc 125:636–637
2. Sandee AJ, Williams CK, Evans NR et al (2004) Solution-processible conjugated electrophosphorescent polymers. J Am Chem Soc 126:7041–7048
3. Evans NR, Devi LS, Mak CSK et al (2006) Triplet energy back transfer in conjugated polymers with pendant phosphorescent iridium complexes. J Am Chem Soc 128:6647–6656
4. Jiang JX, Xu YH, Yang W et al (2006) High-efficiency white-light-emitting devices from a single polymer by mixing singlet and triplet emission. Adv Mater 18:1769–1773
5. Wu F-I, Yang X-H, Neher D et al (2007) Efficient white-electrophosphorescent devices based on a single polyfluorene copolymer. Adv Funct Mater 17:1085–1092
6. Chien CH, Liao SF, Wu CH et al (2008) Electrophosphorescent polyfluorenes containing osmium complexes in the conjugated backbone. Adv Funct Mater 18:1430–1439
7. Chen YC, Huang GS, Hsiao CC et al (2006) High triplet energy polymer as host for electrophosphorescence with high efficiency. J Am Chem Soc 128:8549–8558
8. Wu ZL, Xiong Y, Zou JH et al (2008) High-Triplet-Energy poly 9,9 ‘-bis(2-ethylihexyl)-3,6-fluorene) as host for blue and green phosphorescent complexes. Adv Mater 20:2359–2364
9. Liu J, Pei Q (2010) Poly(M-Phenylene): conjugated polymer host with high triplet energy for efficient blue electrophosphorescence. Macromolecules 43:9608
10. Shao KF, Xu XJ, Yu G et al (2006) Blue electrophosphorescent light-emitting device using a novel nonconjugated polymer as host materials. Chem Lett 35:404–405
11. Yeh HC, Chien CH, Shih PI et al (2008) Polymers derived from 3,6-fluorene and tetraphenylsilane derivatives: solution-processable host materials for green phosphorescent oleds. Macromolecules 41:3801–3807
12. Huang SP, Jen TH, Chen YC et al (2008) Effective shielding of triplet energy transfer to conjugated polymer by its dense side chains from phosphor dopant for highly efficient electrophosphorescence. J Am Chem Soc 130:4699–4707
13. King SM, Al-Attar HA, Evans RJ et al (2006) The use of substituted iridium complexes in doped polymer electrophosphorescent devices: the influence of triplet transfer and other factors on enhancing device performance. Adv Funct Mater 16:1043–1050
14. Shao S, Ma Z, Ding J et al (2012) Spiro-linked hyperbranched architecture in electrophosphorescent conjugated polymers for tailoring triplet energy back transfer. Adv Mater 24:2009–2013
15. Dexter DL (1953) A theory of sensitized luminescence in solids. J Chem Phys 21:836–850
16. Saragi TPI, Spehr T, Siebert A et al (2007) Spiro compounds for organic optoelectronics. Chem Rev 107:1011–1065
17. Ostrowski JC, Robinson MR, Heeger AJ et al. (2002) Amorphous iridium complexes for electrophosphorescent light emitting devices. Chem Commun 7:784–785
18. Yu XM, Zhou GJ, Lam CS et al (2008) A yellow-emitting iridium complex for use in phosphorescent multiple-emissive-layer white organic light-emitting diodes with high color quality and efficiency. J Organomet Chem 693:1518–1527
19. Sudhakar M, Djurovich PI, Hogen-Esch TE et al (2003) Phosphorescence quenching by conjugated polymers. J Am Chem Soc 125:7796–7797
20. McGee KA, Mann KR (2007) Selective low-temperature syntheses of facial and meridional tris-cyclometalated iridium(III) complexes. Inorg Chem 46:7800–7809
21. Wong KT, Liao YL, Lin YT et al (2005) Spiro-configured bifluorenes: highly efficient emitter for uv organic light-emitting device and host material for red electrophosphorescence. Org Lett 7:5131–5134

22. Gong X, Ostrowski JC, Bazan GC et al (2003) Electrophosphorescence from a conjugated copolymer doped with an iridium complex: high brightness and improved operational stability. Adv Mater 15:45–49
23. Gong X, Ostrowski JC, Moses D et al (2003) Electrophosphorescence from a polymer guest-host system with an iridium complex as guest: forster energy transfer and charge trapping. Adv Funct Mater 13:439–444

Chapter 6
Conclusions and Outlook

In conclusion, this thesis outlines the invention of a series of novel, bipolar partially conjugated polyarylether hosts containing triphenylphosphine oxide/carbazole as electron-/hole-transport units. The high E_Ts of these hosts (up to 2.96 eV) have broken through the limitations of traditional conjugated polymer hosts. The significance of this breakthrough is that the long-standing bottleneck problem of triplet energy back transfer from dopants to hosts in blue PhPLEDs is overcome, and thus high efficiency of 23.3 cd/A can be achieved for these devices. As a consequence of this achievement, the design of high-efficiency blue electrophosphorescent polymers and all-phosphorescent white ones, which was unachievable before, becomes available in this work, with state-of-the-art efficiencies of 19.4 and 18.4 cd/A achieved for blue and white PhPs, respectively. In addition, the author also describes a spiro-linked hyperbranched architecture for PhPs to inhibit undesired triplet energy back transfer without high triplet energy hosts, which provides us a new insight into the nature of the triplet energy transfer process.

Despite these achievements, significant progress is still needed before these polymers become commercially viable. First, for polymer hosts, further exploring their physical and chemical properties is essential for device performance optimization. Tailoring their chemical structures to tune the E_Ts within a wide range to meet the requirements of dopants with different colors is of significance for full color display. In particular, universal hosts with suitable E_Ts for all blue, green, and red dopants are necessary for developing white eletrophophorescent polymer systems. Moreover, HOMO/LUMO levels of the hosts should be finely modulated to match well the adjacent layers to further reduce the charge injection barriers for achieving high power efficiency. Secondly, for electrophosphorescent polymers, phosphorescent dopants with suitable emission wavelength, high photoluminescence quantum efficiency, as well as short triplet lifetime should also be searched. Thirdly, polymerization methodology is another important issue for these polymers. Phosphorscent complexes are not durable under many harsh conditions, so mild polymerization conditions are superior for the successful synthesis of these polymers. Fourthly, auxiliary materials including high E_T hole/electron injection/transporting

S. Shao, *Electrophosphorescent Polymers Based on Polyarylether Hosts*,
Springer Theses, DOI 10.1007/978-3-662-44376-7_6

polymers are needed to precisely control the exciton recombination zone in the emissive layer. This is challenging because these polymers must be compatible with the solution-processed multiple-layer stacking structure of PhPLEDs. Finally, lifetimes and reliabilities of PhPLEDs based on electrophosphorescent polymers are critically important issues for their commercialization, which have not been well addressed yet. In general, high EL efficiency, various emission colors, good color purity, and long operational lifetime are basic requirements for these materials. We believe that, with continuous efforts of both the academic and industrial communities, display and lighting products based on electrophosphorescent polymers will be practically applicable in the future.

Zeitfracht Medien GmbH
Ferdinand-Jühlke-Straße 7
99095 Erfurt, Deutschland
produktsicherheit@kolibri360.de